THE ROMANCE OF PLANT LIFE

INTERESTING DESCRIPTIONS OF THE STRANGE AND CURIOUS IN THE PLANT WORLD

G. F. SCOTT ELLIOT

Édition et mise en page : Culturea (Hérault, 34)
Retrouvez notre catalogue sur http://culturea.fr/
Contact : infos@culturea.fr
Imprimé en Allemagne par Books on Demand Gmbh
Design typographique et layout : Derek Murphy et Reedsy
ISBN : 9791041985579
Date de parution : février 2024

CHAPTER I. THE ACTIVITY OF VEGETABLES.

PLANTS WHICH MOVE—SENSITIVE PLANT—A TOURIST FROM NEPTUNE—THE WORLD'S AND THE BRITISH HARVEST—WORKING OF GREEN LEAVES—POWER OF SUNSHINE—WORK DONE BY AN ACRE OF PLANTS—COLTSFOOT, DANDELION, PANSIES, IN SUNSHINE AND IN COLD—WOODSORREL AND CROCUS—FOXGLOVE—LEAVES AND LIGHT—ADVENTURES OF A CARBON ATOM—THE SAP—CABBAGES AND OAKS REQUIRING WATER—TRAVELLER'S TREE—THE WATER IN TREES—AN OASIS IN GREECE—THE ASSOCIATE LIFE OF ITS TREES AND FLOWERS.

WHEN we remember either the general appearance or the way in which a cabbage or a turnip appears to exist, it does not seem possible to call them active. It is difficult to imagine anything less lively than an ordinary vegetable. They seem to us the very model of dullness, stupidity, and slowness; they cannot move even from one field to the next; they are "fast rooted in the soil"; "they languidly adjust their vapid vegetable loves" like Tennyson's Oak.

In fact one usually speaks of vegetating when anybody is living a particularly dull, unexciting kind of life in one particular place.

And it even seems as if the books, which are supposed to give us the best information about the study of plants, and which are not very attractive little books, quite agree with the ordinary views of the subject.

For one finds in them that plants differ from animals in being "incapable of motion." This, of course, just means that an animal, or rather most animals, can walk, swim, or fly about, whilst plants have roots and do not move from one spot to another. But it is not true to say that plants cannot move, for most plants grow, which means that they move, and in some few cases, we find that plants behave very much in the same way as animals do when they are touched or excited in any way.

We shall have to speak about tendrils, roots, and insect-catching plants later on. But it is perhaps the Sensitive Plant which shows most distinctly that it can shrink back or shrink together when it is bruised or roughly handled.

It will be described in its place, but just to show that this plant can move of its own accord, it is only necessary to hold a lighted or burning match about an inch or so below the end of a long leaf. If one does this then all the little leaflets begin to fold up, and finally the main stalk droops; soon afterwards other leaves higher up the stalk begin to be affected in the same way, and fall limply down one after the other. It is supposed that this movement frightens a grazing animal, who will imagine there is something uncanny about the plant and leave it alone. There are many respects in which this reaction of the Sensitive Plant resembles that found in animals. It does not take place if the plant is chloroformed or treated with ether; the leaves also get "fatigued" if too often handled, and refuse to rise up again.

There are, however, only a very few plants in which an immediate, visible answer to a stimulus can be detected. But all plants are at work; they have periods of rest which correspond to our sleep, but during their ordinary working hours they never slacken off, but continue vigorously active.

The life of man is so short that it is difficult to realize all that is being done by the world of plants. It is necessary to get beyond our human ideas of time. That is most conveniently done by considering how our plant world would strike an inhabitant of the planet Neptune. Our theoretical Neptunian would be accustomed to a year of 60,127 days (164 of our years); we will suppose that three of our years are a Neptunian week, and that ten of our days are about three-quarters of a Neptunian hour, whilst two earth-hours would be a minute to him.

If such a being were to observe our earth, he would be astonished at the rapidity of our vegetable world. The buds would seem to him to swell visibly; in the course of an hour or two, the bare boughs of the trees would clothe themselves with the luxuriant greenery of midsummer. Hops would fly round and round their poles, climbing at the rate of a foot a minute. Bare places, such as the gravel heaps near a sandpit, or the bare railroad tracks at a siding, would be perhaps in one week entirely covered by rich grass and wild flowers. In six Neptunian months a forest of graceful larches would spring up to a height of seventy or eighty feet.

So that, if one thinks Neptunially, the activity of plants can be easily realized.

The truth is that we are so familiar with common annual events, such as the regular harvest every year, that we never seem to realize what it means. There are some 1,400,000,000 human beings on the earth to-day, and they entirely depend on the work done every year by cultivated and wild plants.

Even in one of the least agricultural of all civilized countries, such as Great Britain, the cultivation of plants is still the largest national industry. In 1897 we grew enough corn to give a ration of lb. per diem to every inhabitant for 68 days, and we manage to get a large amount from every acre (28 to 33 bushels per acre). In most other countries the relative importance of land and of agriculture generally is very much greater than it is in Britain.

Moreover, it seems at first sight as if all this harvest had been made out of nothing at all. Plants do take in a small amount of mineral matter from the earth, but these minerals form but a very little part of the bulk of a tree or any vegetable substance.

A piece of wood can be burnt up in a fire and very little indeed of it is left. A few ashes will indeed remain, which are the minerals taken in from the earth, but all the rest has vanished into the atmosphere. The water which was contained in the wood has become steam and is evaporated; the woody matter consisted chiefly of compounds of a chemical substance, carbon, which also becomes an invisible gas (carbonic acid gas) in a fire and goes back into the atmosphere.

When the piece of wood was formed in a growing tree, it is easy to see where the water came from: it was taken in by the roots. Just as flowers drink up the water in a vase, and wither if they do not receive enough, so all plants suck up water by their roots. The carbonic acid gas is taken into plants through their leaves and is worked up into sugar, starch, wood, and other matters inside the plant.

But there is another very interesting point about the way in which wood is burnt in a fire; heat and light are obtained from a wood fire. Where did that heat and light come from?

If you walk in summer, under a tree in full leaf, it is much cooler than it is in the sunshine outside. This shows what happens: the sunshine has been taken up or absorbed by the leaves of the tree. It does not pass through the foliage, but the heat and light are stopped by the leaves.

The light and heat which were used up by the leaves in making wood, sugar, and starch come back again when that wood or starch is burnt.

So that the burning up of a bit of wood is just the opposite to the formation of that wood in sunshine in a living tree. The important point is that it is the sunshine which is used by plants to make all these refractory bodies, such as water, carbonic acid gas, and others, unite together to form sugar, starch, and wood.

As the earth revolves upon its axis, sunlight falls successively on every acre of land. Almost everywhere it is intercepted by green foliage. Each leaf of every plant receives and absorbs as much as it can, and, for so long as the light lasts, its living particles are hard at work: water or sap is hurrying up the stem and streaming out of the leaves as water vapour. Carbonic acid gas also is hurrying into the leaves; inside these latter first sugar and then starch is being manufactured, so that the green cells become filled with starch or sugar.

So soon as the light fails, the work begins to slacken. When darkness sets in, the starch changes to sugar and passes down the leaf-stalk into the stem, where it is used up in growth, in the formation of new wood or in supplying the developing flowers or young buds.

Next morning when the sunlight touches the plant all its little living cells set to work again, and another day's task is begun. It is very difficult to understand what is going on inside the leaf. If you were to imagine a square yard of leaves all taking in sunshine and making starch as they do in fine weather; then if you weighed all these leaves, and then weighed them again one hour after they had been in the sunshine, of course that square yard of leaf surface should be heavier, because a certain amount of starch has been formed in it. The amount actually made in one hour has been estimated by Dr. Horace Brown as 1/500 lb. So that 100 square yards of leaves working in sunshine for five hours might make one pound of starch. But one can estimate the activity of plants in another way. Look at the amount of work done by the Grass, etc., on an acre of pasture land in one year. This might entirely support a cow and calf during the summer; all the work done by these animals, as well as all the work which can be done on the beef which they put on, is due to the activity of the grasses on that acre. Moreover it is not only these large animals that are supported, but every mouse, every bird, every insect, and every worm which lives on that piece of ground, derives all its energy from the activity of the plants thereon.

All work which we do with our brains or muscles involves the consumption of food which has been formed by plants under the warm rays of the sun.

So that man's thoughts and labour, as well as that of every living creature, is in the first instance rendered possible by sunshine.

But the sunlight, besides this all-important function, affects plants in other ways.

One of the most interesting of the early spring flowers is the Coltsfoot. On bare blackish and unsightly heaps of shale one may see quantities of its golden blossoms. Now if one looks at them on a fine sunny day, every single blossom will be widely opened and each will turn towards the sun.

In wet cold weather every blossom will hang its head and be tightly closed up. Exactly the same may be observed with the Dandelion, which is, indeed, still more sensitive than the Coltsfoot. In cold wet weather it is so tightly closed that it is barely possible to make out the yellow colour of the flower, but on warm sunny days it opens wide: every one of its florets drinks in as much as possible of the genial sunshine. Both opening and closing are produced by the warmth and light of the sun's rays.

It is also the same with Pansies. On a fine day they spread out widely, but in cold wet weather the heads hang over and the whole flower shrinks together.

Perhaps the most interesting of them all are the little Woodsorrel and the Crocus.

Both are exceedingly sensitive to sunlight, or rather to the cold. A mere cloud passing over the sun on a fine spring morning will close up the flowers of the Crocus. In cold weather, if you bring one of its flowers indoors and put it near a bright light it will open widely, sometimes in a few minutes.

What produces these changes? It is very difficult to say, but every change helps towards the general good of the plant. In warm sunny weather insects are flying about, and they can enter the flower if it is open. These insects help in setting the seed (as we shall see in another chapter). In cold wet weather the flowers are best closed, as the rain might injure the florets and because also no insects are abroad.

Both the Foxglove and the Blue Vetch (*Vicia Cracca*) are specially ingenious in their way of obtaining light. For the stalk of every separate blossom bends so that its head turns to the best lighted or sunniest side. Thus, if you have Foxgloves planted against a wall, every flower will turn away from it; if you plant them in a circular bed, every one turns to the outside, so that every flower can get the sunlight.

Every one who has kept plants in a window knows that the stems turn towards the light. This has the effect of placing the leaves where they can get as much sunshine as possible. The leaves themselves are also affected by sunlight. They seem to stretch out in such a way that they absorb as much of it as they can.

That, of course, is what they ought to do, for they want to obtain as much as possible of the sunlight to carry on the work of forming sugar and starch inside the leaf.

Not only each leaf by itself endeavours to place itself in the best light-position, but all the leaves on the same spray of, for instance, Elm, Lime, or Horsechestnut, arrange themselves so that they interfere with one another as little as possible. Very little light is lost by escaping between the leaves, and very few of the leaves are overshaded by their neighbours on the same branch.

Thus all co-operate in sunlight-catching. But, when a number of different plants are competing together to catch the light on one square yard of ground, their leaves try to overreach and get beyond their neighbours.

On such a square yard of ground, it is just the competition amongst the plants, that makes it certain that every gleam of light is used by one or other of them.

Every one of all those plants of itself alters the slope of its leaves and turns its stems so as to get as much light as possible.

This light, as we have seen, is taken in by the plant. It is used to make the gas, carbonic acid, unite with water: when these are made to join together, they form sugar; if the sugar is burnt the heat and light appear again.

By changing the amount and arrangement of the molecules in sugar, starch or vegetable fats, and many other substances can be formed. But it is the sunlight that makes all this possible.

Thus the sun not merely supplies the motive power for all animal and vegetable activity but, by its influence, flowers, leaves, and stems move and turn in such ways that they are in the most convenient position to intercept its light.

The sunlight, though all-important in the life of most plants, kills many kinds of bacteria and bacilli which love the darkness. The well-known radium rays are also destructive to bacteria, and hinder the growth of certain fungi (Becquerel's rays have a similar effect). The X-rays are not so well understood, but one can close the leaflets of the Sensitive Plant by means of them.

Carbonic acid gas forms but a small proportion of the atmosphere which surrounds a growing plant. Yet there is no lack of it, for when the leaf is at work forming sugar the particles of gas are rushing into the leaf, and other particles come from elsewhere to take their place. Every fire and every breath given off by an animal yields up carbonic acid, so that it is constantly in circulation.

This is more easily seen by tracing the probable history of an atom of carbon. We will suppose that it enters a grass leaf as carbonic acid gas and becomes starch: next evening it will become sugar and may pass from cell to cell up the stem to where the fruit or grain is ripening. It will be stored up as starch in the grain. This grass will become hay and in due course be eaten by a bullock. The starch is changed and may be stored up in the fat of the animal's body. When this is eaten at somebody's dinner, the fat will most probably be consumed or broken up; this breaking up may be compared to a fire, for heat is given off, and the heat in this case will keep up the body-temperature of the person. The carbon atom will again become carbonic acid gas, for it will take part of the oxygen breathed in, and be returned to the atmosphere as carbonic acid gas when the person is breathing.

Another atom of carbon might enter the leaves of a tree: it will be sent down as sugar into the trunk and perhaps stored up as vegetable fat for the winter. Next spring the vegetable fat becomes starch and then sugar: as sugar it will go to assist in forming woody material. It may remain as wood for a very long time, possibly 150 to 200 years: then the tree falls and its wood begins to decay.

The bark begins to break and split because beetles and woodlice and centipedes are burrowing between the bark and the wood. Soon a very minute spore of a fungus will somehow be carried inside the bark, very likely sticking to the legs of a beetle. This will germinate and begin to give out dissolving ferments which, with the aid of bacteria, attack the wood. Our carbon atom is probably absorbed into the fungus. Very soon the mushroom-like heads of this fungus begin to swell and elongate; they burst through the bark and form a clump of reddish-yellow Paddock-stools. A fly comes to the fungus and lays an egg in it. This egg becomes a fat, unpleasant little maggot which eats the fungus, and amongst others devours our carbon atom, which again becomes fat in its body. Then a tomtit or other small bird comes along and eats the maggot. That bird stays out too late one evening and is eaten by an owl. The owl, satisfied with a good meal, allows itself to be surprised and shot by a keeper. When its body is nailed to a door and decays away, the carbon atom again takes up oxygen and becomes carbonic acid gas, which escapes into the atmosphere, and is ready for a fresh series of adventures.

We must now consider the water which with carbonic acid gas makes up sugar, etc. All plants contain a large percentage of water. This may be as much as 95 to 98 per cent in water plants, and 50 to 70 per cent. in ordinary tissues; it is contained in every sort of vegetable substance.

But there is also a stream of water or sap which is almost always entering the roots, rising up the stem, and passing into the leaves. On these leaves there are hundreds of minute openings called stomata, by which the water escapes as water-vapour into the atmosphere. A single oak leaf may have 2,000,000 of these stomata.

It is this current of sap which keeps the leaf fresh and vigorous; it is also by this current that every living cell is supplied with water and kept in a strong, healthy condition.

The amount of water used in this way is very great; in four months an acre of cabbages will transpire or give out through its leaves 3,500,000 pints of water and an acre of hops from 5-1/2 to 7 millions. A single oak tree,4 supposed to have 700,000 leaves, must apparently have given off into the atmosphere during five months 230,000 lb. of water.

Sometimes the water is so abundant in the plant that it collects as drops on the tips of the leaves and falls off as fluid water. A very young greenhouse plant (*Caladium nymphaefolium*) was found by Molisch to give off water-drops a minute, and in one night it exuded one-seventeenth of a pint.

The water is found stored up in the stems or leaves of plants, especially those of hot or dry climates. The Madagascar Traveller's Tree, *Ravenala,* has a considerable amount of water in a hollow at the base of its leaf, and it is possible to drink this water. The usual story is to the effect that a panting traveller finds this palm in the middle of the desert, and saves his life by quenching his thirst with its crystal-clear water. Unfortunately the tree never grows far from marshy ground or springs, and the water, which I tasted for curiosity, had an unpleasant vegetable taste, with reminiscences of bygone insect life.

These are, of course, exceptional cases; as a rule the tiny root-hairs search and explore the soil; the sap or ascending current passes up the stem and pours out into the atmosphere. There the vapour is hurried off by winds, and eventually condenses and, falling as snow or rain on the earth, again sinks down into the soil.

It is very difficult to understand how the sap or water rises in the trunks of tall trees; we know that along the path of the sap inside, the root-hairs and other cells in the root, the various cells in the stem, and finally those of the leaf, are all kept supplied and distended or swollen out with water. All these living cells seem to have the power of absorbing or sucking in water, and eventually they are so full and distended within, that the internal pressure becomes almost incredible. Wieler found in the young wood of a Scotch fir that the pressure was sixteen atmospheres, or 240 lb. to the square inch. Dixon, when experimenting with leaf-cells, found ten, twenty, or even thirty atmospheres (150 to 450 lb. to the square inch). No locomotive engine has cylinders strong enough to resist such internal pressures as these. It is an extraordinary fact, and one almost incredible, that the cells can stand such pressures.

This and such palms are often placed at the mouths of cañons to indicate water, and may, indeed, thus save the lives of passing travellers.

Yet these minute living cells not only exist but work at this high tension, and, in some cases, they live to about fifty years.

In this favoured country of Great Britain, it is unusual to find any serious lack of water. But in Italy or Greece, every drop of it is valuable and carefully husbanded.

Sometimes in such arid dry countries, a small spring of water will form around itself a refreshing oasis of greenery surrounded everywhere by dreary thorn-scrub or monotonous sand. All the plants in such a spot have their own special work to do: the graceful trees which shade the spring, the green mosses on the stones, the fresh grass and bright flowers or waving reeds, are all associated in a common work. They protect and shelter each other; their dead leaves are used to form soil; their roots explore and break up the ground. It is true that they are competing with one another for water and for light, but they are all forming a mutual protection, and producing an annual harvest.

In a climate like our own we cannot, like the Greek, suppose a Nymph in the shape of a lovely young woman watching over the spring, for she would infallibly suffer from rheumatism and ague.

But every living cell in every plant in such an oasis depends upon the water of the spring. All the plants there form an association which can be quite well compared to a city or some other association of human beings. They do compete, for they struggle to do the most work for the good of the community, and they incidentally obtain their livelihood in the process.

Most plant societies or associations such as those which cover Great Britain are not so obviously dependent on one particular spring, but the plants composing them are associated in a very similar way.

CHAPTER II. ON SAVAGES, DOCTORS, AND PLANTS.

SAVAGES KNEW BOTANY—FIRST LADY DOCTORS AND BOTANICAL EXCURSIONS—TRUE DRUGS AND HORRIBLE ORNAMENTS—HYDROPHOBIA CURE—CLOVES—MUSTARD—IVY—ROSES AND TEETH—HOW TO KEEP HAIR ON—HOW TO KNOW IF A PATIENT WILL RECOVER—CURIOUS PROPERTIES OF A MUSHROOM—THE SCYTHIAN LAMB—QUININE: HISTORY AND USE—SAFFLOWER—ROMANCE OF IPECACUANHA—WARS OF THE SPICE TRADE—CINNAMON, DOGWOOD, AND INDIGO—ROMANCE OF PEPPER—BABYLONIAN AND EGYPTIAN BOTANISTS—CHINESE DISCOVERIES—THEOPHRASTUS—MEDIEVAL TIMES—THE FIRST ILLUSTRATED BOOK—NUMBERS OF PLANTS KNOWN—DISCOVERIES OF PAINTERS AND POETS.

IF we look back to the time when all men and women were mere savages, living like the Esquimaux or the Australians of to-day, then it is certain that every person was much interested in plants. Nothing was so interesting as daily food, because no one was ever certain of even one good meal in the day.

So that in those early times there was a very sound, well-grounded knowledge of roots, bulbs, and fruits. They knew all that were good to eat, all that could possibly be eaten in time of famine and starvation, and also every poisonous and unwholesome plant.

Some savage genius must have discovered that certain plants were "good medicine"; that certain tree-barks helped to check fever, and that others were worth trying when people had successfully devoured more than they could comfortably digest. The life of a savage meant tremendous meals, followed by days of starvation; even now, when young children are fed on rice in India, a thread is tied round their waist, and, when this bursts, they are not allowed to eat any more.

Very probably some of these early physicians were lady doctors usually of a certain age. Men were too busy with their hunting and warfare to have time to try experiments with drugs, to make concoctions of herbs all more or less disquieting and to find out if these were of any use.

So that such medicine-men or witches gradually came to understand enough about poisons or fruits to make themselves respected and even feared. They would, no doubt, make botanical excursions in the forest, accompanied by their pupils, in order to point out the poisonous and useful drugs.

It is worth noting, in passing, that this habit of botanical professors going on excursions with medical students has persisted down to our own times, probably without any break in the continuity.

But it was soon found advisable to make this knowledge secret and difficult to get. They did not really know so very much, and a mysterious, solemn manner and a quantity of horrible and unusual objects placed about the hut would perhaps prevent some irate and impatient savage patient from throwing a spear at his wizard—or witch-doctor.

Shakespeare alludes to this in *Macbeth*. "Scale of Dragon; tooth of wolf; witches' mummy; maw and gulf of the ravin'd salt sea shark; root of hemlock digg'd i' the dark; ... gall of goat and slips of yew"; and so on.

Most of their cures were faith-cures, and they were, no doubt, much more likely to be successful when the patient believed he was being treated with some dreadful stew of all sorts of wonderful and horrible materials.

This explains how it was that the knowledge of medicine became so mixed up with pure charlatanism and swindling that no man could tell which drugs were of real use and which were mere ornaments giving piquancy and flavour to the prescription. It is not possible to say that a snake's head, the brain of a toad, the gall of a crocodile, and the whiskers of a tiger, were all of them absolutely useless. Within the last few years it has been found that an antidote to snake-bite can be obtained from a decoction of part of the snake itself, and it has also been discovered that small quantities of virulent poisons are amongst our most valuable and powerful remedies.

Whether the savages and their successors the doctors of feudal times even down to the fifteenth and sixteenth centuries, suspected or believed that this was the case must remain a rather doubtful hypothesis, but there is no question "that the hair of the dog that bit him" theory of medicine was very prevalent.

The following was a cure for hydrophobia of a more elaborate nature: "I learned of a Friend who had tried it effectual to cure the Biting of a Mad Dog; take the Leaves and Roots of Cowslips, of the leaves of Box and Pennyroyal of each a like quantity; shred them small to put them into Hot Broth and let it be so taken Three Days Together and apply the herbs to the bitten place with Soap and Hog's suet melted together" (Parkinson).

This prescription is not so preposterous as it sounds. Box and Pennyroyal both contain essences which would be in all probability fatal to the germ of hydrophobia, and the soap and hog's suet would keep air from the wound.

Other prescriptions read like our modern patent medicines.

"Good Cloves comfort the Brain and the Virtue of Feeling, and help also against Indigestion and Ache of the Stomach" (Bartholomew).

"Senvey" (the old name of mustard) "healeth smiting of Serpents and overcometh venom of the Scorpions and abateth Toothache and cleanseth the Hair and letteth" (that is, prevents or tends to prevent) "the falling thereof. If it be drunk fasting, it makes the Intellect good."

Even in those days the people can scarcely have believed that drinking mustard improved the intellect. Many of the remedies and cures are obviously false, for example the following:-

"A man crowned with Ivy cannot get drunk."

"Powder of dry Roses comforteth wagging Teeth that be in point to fall."

The fact that the surgeon was also a barber, and also a "face-specialist," appears from the two following: —

"Leaves of Chestnut burnt to powder and tempered with Vinegar and laid to a man's Head plaisterwise maketh Hair increase and keepeth hair from falling."

Those whose hair turned grey could employ the following prescription: —

"Leaves of Mulberry sod in rainwater maketh black hair."

If a doctor was not quite sure of the endurance of a patient under these heroic remedies, he could easily find out if he would recover, for it was only necessary to try the following:—

"Celandine with the heart of a Mouldwarp" (that is mole, *Scottice* moudiewort) "laid under the Heade of one that is grievouslie Sicke, if he be in danger of Death, immediately he will cry out with a loud voice or sing; if not, he will weep."

In Lightfoot's *Flora Scotica*, there is an interesting account of the Fly Mushroom (*Agaricus muscarius*) which is not very rare in Britain, and which may be easily recognized by the bright red top or cap, with whitish scales scattered over it, and a sort of ring of loose white tissue round the stalk.

"It has an acrid and deleterious quality. The inhabitants of Kamschatka prepare a liquor from an infusion of this Agaric which taken in a small quantity exhilarates the spirits, but in a larger dose brings on a trembling of the nerves, intoxication, delirium and melancholy. Linnæus informs us that flies are killed or at least stupefied by an infusion of this fungus in milk and that the expressed juice of it anointed on bedsteads and other places effectually destroys"—what we may describe as certain lively and pertinacious insects with a great affection for man!

As a matter of fact the fungus is said to be a deadly poison.

These quotations are enough to show how the real medical knowledge of those times was encrusted with all sorts of faith-curing devices, sheer falsehoods, and superstitions. The most learned men of the Middle Ages were almost invariably monks and hermits, for there was nothing in the world of those strenuous times to attract a studious, sensitive disposition. The spirit of their learning can be judged from the wearisome disquisitions and lengthy volumes written about the Barnacle Goose and Scythian Lamb.

In certain deserts along the Volga River in Russia, a peculiar fern may be found. It might be described as resembling a gigantic Polypody; the stem is about as thick as a lamb's body and grows horizontally on the ground like that of the common fern mentioned; thick furry scales cover the outside of its stem, which ends at the tip in an elongated point. The blackish-green leaf-stalks springing from the furry stem end in large divided green leaves.

It occurred to some medieval humorist to cut off the upper part of the leaf-stalks, and to make a sort of toy lamb out of the four leaf-stalk stumps and part of the woolly or furry stem.

This was palmed off as a wonderful curiosity of nature, as "a plant that became an animal," upon the ingenuous tourist of the period.

Such a subject was thoroughly congenial to the learned mind in the Middle Ages, and an enormous quantity of literature was produced in consequence. The general theory is given in the following lines:—

"Cradled in snow and fanned by Arctic air,
Shines, gentle Barometz, thy golden hair,
Rooted in earth each cloven hoof descends,
And round and round her flexile neck she bends,
Crops the grey coralmoss and hoary thyme,
Or laps with rosy tongue the melting rime,
Eyes with mute tenderness her distant dam,
Or seems to bleat, a vegetable lamb."

Such is the old idea of a well-known fern, *Cibotium barometz*.

Yet the original researches of some African "Obi" wizard or red Indian were not forgotten, and gradually came into practice.

The title-page of John Parkinson's "Paradisus." In the distance may be seen a Scythian Lamb growing on its tree, and in the foreground many plants are shown as well as Adam and Eve.

It must be remembered that these savages were true scientific experimentalists, and made discoveries which have been of infinite service to mankind. We remember great men like Harvey, Lister, and Pasteur, but we never think of the Indian who discovered quinine.

The quinine trees, the yellow variety or *Calisaya cinchona*, grow in the mountains of north-eastern Bolivia and south-eastern Peru, in wild, inaccessible places at heights of 5000 to 6000 feet. The Indians probably experimented with almost every part of every wild tree before they discovered the wonderful properties of this particular species. The quinine in nature is probably intended to prevent some fungus or small insect from attacking the bark: when quinine is used in malaria, it kills the fever germ which attacks the blood corpuscles of the sick person, so that it is of the utmost importance in all tropical countries.

When the Jesuit fathers reached Peru and made friends and converts of the Indians, they discovered this remedy. Soon after the Countess de Chinchon, wife of the Viceroy of Peru, fell seriously ill of fever and was cured by the use of Jesuit's bark or quinine. It was introduced into Europe about 1638, but for a very long time the entire supply came from South America. The British Indian government were paying some £12,000 every year for South American quinine and, at the same time, the supply was running short, for the Indians were cutting down every tree.

At last, in 1859 (on the suggestion of Dr. Royle in 1839), the adventurous journeys of Clements Markham, Spruce, and Robert Cross resulted in the introduction of the Cinchona now flourishing in Madras, Bombay, and Ceylon. In 1897 34 British colonies produced about £43,415 worth of quinine, and the price is now only 7-1/2d. or 8d. a pound!

Such drugs as Safflower are of very ancient date. It was commonly employed in Egypt with other dyes and spices for embalming mummies. It is now used with carbonate of soda and citric acid to give a pink dye to silks and satins, and occasionally, in the form of rouge, to ladies' cheeks! How did the ancient Egyptians discover that this particular thistle-like plant (*Carthamus tinctorius*) had flowers from which a red dye could be extracted by a tedious process of soaking in water? The natural colour of the flowers is not red but yellow.

The history of other drugs reads like a romance. Ipecacuanha, for instance, was discovered by some unknown Indian who lived in the damp tropical forests of Brazil and New Granada. A worthy merchant in Paris obtained a little of the drug in the way of trade. Shortly afterwards he became very ill and was attended by a certain Dr. Helvetius, who was exceedingly attentive to him. The grateful merchant gave the kind-hearted physician some ipecacuanha. In the course of time the great King Louis XIV's son fell ill of dysentery, and Helvetius received 1000 louis d'or for his ipecacuanha.

A very interesting and romantic history might be written about the effect of drugs, dyes, and spices in developing trade. During the time when Britain was struggling to obtain a share of the foreign trade of Holland and France, such spices as Clove, Cinnamon, and Pepper were of the greatest importance. The Dutch, especially, adopted every possible method to keep the spice trade in their own hands. They cut down the clove, cinnamon, and other trees, in all the islands not directly under their control. They imposed the most barbarous penalties on any interloper. For instance, any one who sold a single stick of cinnamon in Ceylon was punished with death. When the English captured the island in 1796, all such restrictions were of course repealed. Nevertheless its cultivation remained a monopoly of the East India Company until 1832.

Logwood (*Haematoxylon campechianum*) is closely connected with the story of adventure and colonisation in the West Indies. Its use was at first forbidden by Queen Elizabeth as it did not yield fast colours; this was because the dyers of those times did not know of any mordant to fix them. Yet this is one of the few vegetable dyes which retain their position in the market in these days of aniline colours, and it is said to be a large constituent, with brandy, of cheap "port wine."

Indigo was known to the Romans, who imported it from India on camel-back by way of the Persian and Syrian desert. In the fifteenth century, when the Dutch began to introduce it in large quantities, it was found to interfere with the "woad" (*Isatis tinctoria*) which was then a very important cultivated plant in Europe. In Nuremberg, an oath was administered once a year to all the manufacturers and dyers, by which they bound themselves not to use the "devil's dye," as they called Indigo. Its more recent history shows a very different system. In Assam and other parts of British India, enormous sums of money have been invested in indigo plantations. It has been estimated that four million pounds was invested, and that a population of something like 700 Europeans and 850 workmen to the square mile in Behar, were entirely supported by indigo plantations.

Now all these planters are ruined and the population is dispersed, because German indigo manufactured from coal-tar is destroying the sale of the British-grown material. The plant has pretty blue flowers and belongs to the *Leguminous* order. The dye is obtained by steeping the leaves and young branches in water, and it is finally turned out in blue powder or cakes.

Perhaps the most interesting of all these drugs is Pepper. The Dutch, in the days of Queen Elizabeth, had a monopoly of the East Indian trade, and they tried to cut down or burn all spice trees except those in their own control. They could thus form a corner in pepper, and alter the price as they felt inclined. At one period they doubled the price, raising it from three shillings to six shillings per pound. This annoyed the London merchants so much that they met together and formed the "Society of Merchants and Adventurers trading to the East Indies." This was of course the original source of our great East Indian trade, and later on resulted in the Indian Empire.

At present, and for centuries past, the whole world is searched and explored for drugs and spices. Our medicinal rhubarb for instance, grows in China on the frontiers of Tibet; it is carried over the mountains of China to Kiaghta in Siberia, and from thence taken right across Russian Siberia to London and New York. It is closely allied to the common or garden rhubarb, which grows wild on the banks of the Volga.

It is only our duty to remember with gratitude all those long since departed botanists who have made our life so full of luxury and have supplied our doctors with all kinds of medicines.

The first doctors were of course just savage botanists, but as soon as men began to write down their experiences, we find botanical treatises. The first, and for a very long time the only, botanical books were intended to teach medical students the names and how to recognize useful flowers and drugs.

Medicinal herbs such as mandrake, garlic, and mint are found described on those clay cylinders which were used in Babylon instead of books, about 4000 B.C., that is some 6000 years ago! The Egyptians thought that "kindly, healing plants," such as opium, almonds, figs, castor-oil, dates, and olives, were derived from the "blood and tears of the gods"; that would be about 3000 B.C. It is not known how far back Chinese botany can be traced, but, by the twelfth century before Christ, some three hundred plants were known, including ginger, liquorice, rhubarb, and cinnamon.

Theophrastus, who flourished about 300 B.C., was a scientific botanist far ahead of his time. His notes about the mangroves in the Persian gulf are still of some importance. It is said that some two thousand botanical students attended his lectures. It is doubtful if any professor of botany has ever since that time had so large a number of pupils. Dioscorides, who lived about 64 B.C., wrote a book which was copied by the Pliny (78 A.D.), who perished in the eruption of Vesuvius. The botany of the Middle Ages seems to have been mainly that of Theophrastus and Dioscorides. In the tenth century we find an Arab, Ibn Sina, whose name has been commemorated in the name of a plant, Avicennia, publishing the first illustrated text-book, for he gave coloured diagrams to his pupils.

After this there was exceedingly little discovery until comparatively recent times.

But Grew in 1682 and Malpighi in 1700 began to work with the microscope, and with the work of Linnæus in 1731 modern botany was well started and ready to develop.

It is interesting to compare the numbers of plants known at various periods, so as to see how greatly our knowledge has been increased of recent years. Theophrastus (300 B.C.) knew about 500 plants. Pliny (78 A.D.) knew 1000 species by name. Linnæus in 1731 raised the number to 10,000. Saccardo in 1892 gives the number of plants then known as follows:—

Flowering Plants	105,231	species
Ferns	2819	"
Horsetails and Club-mosses	565	"
Mosses	4609	"
Liverworts	3041	"

Lichens	5600	"
Fungi	39,663	"
Seaweeds	12,178	"

	173,706	

But, during the years that have elapsed since 1892, many new species have been described, so that we may estimate that at least 200,000 species are now known to mankind.

But it is in the inner meaning and general knowledge of the life of plants that modern botany has made the most extraordinary progress. It is true that we are still burdened with medieval terminology. There are such names as "galbulus," "amphisarca," and "inferior drupaceous pseudocarps," but these are probably disappearing.

The great ideas that plants are living beings, that every detail in their structure has a meaning in their life, and that all plants are more or less distant cousins descended from a common ancestor, have had extraordinary influence in overthrowing the unintelligent pedantry so prevalent until 1875.

Yet there were many, not always botanists, of much older date, who made great discoveries in the science. Leonardo da Vinci, the great painter, seems to have had quite a definite idea of the growth of trees, for he found out that the annual rings on a tree-stem are thin on the northern and thick on the southern side of the trunk. Dante seems to have also understood the effect of sunlight in ripening the vine and producing the growth of plants (*Purgatorio*, xxv. 77). Goethe seems to have been almost the first to understand how leaves can be changed in appearance when they are intended to act in a different way. Petals, stamens, as well as some tendrils and spines, are all modified leaves. There is also a passage in Virgil, or perhaps more distinctly in Cato, which is held to show that the ancients knew that the group of plants, *Leguminosæ*, in some way improved the soil. I have also tried to show that Shelley had a more or less distinct idea of the "warning" or conspicuous colours (reds, purples, spotted, and speckled) which are characteristic of many poisonous plants.

But if we begin with the unlettered savage, one can trace the very slow and gradual growth of the science of plant-life persisting all through the Dark Ages, the Middle Ages, and recent times, until about fifty or sixty years ago, when a sudden great development began, which gives us, we hope, the promise of still more wonderful discoveries.

CHAPTER III. A TREE'S PERILOUS LIFE.

HEMLOCK SPRUCE AND PINE FORESTS—STORY OF A PINE SEEDLING—ITS STRUGGLES AND DANGERS—THE GARDENER'S BOOT—TURPENTINE OF PINES—THE GIANT SAWFLY—BARK BEETLES—THEIR EFFECT ON MUSIC—STORM AND STRENGTH OF TREES—TALL TREES AND LONG SEAWEEDS—EUCALYPTUS, BIG TREES—AGE OF TREES—VENERABLE SEQUOIAS, OAKS, CHESTNUTS, AND OLIVES—BAOBAB AND DRAGONTREE—RABBITS AS WOODCUTTERS—FIRE AS PROTECTION—SACRED FIRES—DUG-OUT AND BIRCH-BARK CANOES—LAKE DWELLINGS—GRAZING ANIMALS AND FOREST DESTRUCTION—FIRST KIND OF CULTIVATION—OLD FORESTS IN ENGLAND AND SCOTLAND—GAME PRESERVING.

"The murmuring pines and the hemlocks
Stand like harpers hoar with beards that rest on their bosom."—*Longfellow.*

OF course the Hemlock here alluded to is not the "hemlock rank growing on the weedy bank," which the cow is adjured not to eat in Wordsworth's well-known lines. (If the animal had, however, obeyed the poet's wishes and eaten "mellow cowslips," it would probably have been seriously ill.) The "Hemlock" is the Hemlock spruce, a fine handsome tree which is common in the forests of Eastern North America.

These primeval forests of Pine and Fir and Spruce have always taken the fancy of poets. They are found covering craggy and almost inaccessible mountain valleys; even a tourist travelling by train cannot but be impressed by their sombre, gloomy monotony, by their obstinacy in growing on rocky precipices on the worst possible soil, in spite of storm and snow.

This species of fir tree grows to an enormous height in British Columbia. It is now being planted in many Scotch forests.

But to realize the romance of a Pine forest, it is necessary to tramp, as in Germany one sometimes has to do, for thirty miles through one unending black forest of Coniferous trees; there are no towns, scarcely a village or a forester's hut. The ground is covered with brown, dead needles, on which scarcely even green moss can manage to live.

Then one realizes the irritating monotony of the branches of Pines and Spruces, and their sombre, dark green foliage produces a morose depression of spirit.

The Conifers are, amongst trees, like those hard-set, gloomy, and determined Northern races whose life is one long, continuous strain of incessant endeavour to keep alive under the most difficult conditions.

From its very earliest infancy a young Pine has a very hard time. The Pine-cones remain on the tree for two years. The seeds inside are slowly maturing all this while, and the cone-scales are so welded or soldered together by resin and turpentine that no animal could possibly injure them. How thorough is the protection thus afforded to the young seeds, can only be understood if one takes a one-year-old unopened cone of the Scotch Fir and tries to get them out. It does not matter what is used; it may be a saw, a chisel, a hammer, or an axe: the little elastic, woody, turpentiny thing can only be split open with an infinite amount of trouble and a serious loss of calm.

When these two years have elapsed, the stalk of the cone grows so that the scales are separated, and the seeds become rapidly dry and are carried away by the wind.

These seeds are most beautiful and exquisitely fashioned.

The seed itself is small and flattened. It contains both resin and food material, and is enclosed in a tough leathery skin which is carried out beyond the seed into a long, very thin, papery wing, which has very nearly the exact shape of the screw or propeller of a steamer. This wing or screw is intended to give the seed as long a flight in the air as possible before it reaches the ground. If you watch them falling from the tree, or throw one up into the air and observe it attentively, you will see that it twirls or revolves round and round exactly like the screw of a steamship. It is difficult to explain what happens without rather advanced mathematics, but it is just the reverse of what happens in the steamer.

The machinery in the steamer turns the screw, and the pressure of the water, which is thrown off, forces the boat through the water; in the case of the pineseed, the pressure of the air on the flying wings makes the seed twirl or turn round and round, and so the seed must be a much longer time in falling. They often fly to about 80 or 100 yards away from the parent tree.

Once upon the ground, the seed has to germinate *if it can*; its root has to pierce the soil or find a way in between crevices of rocks or sharp-edged stones. All the time it is exposed to danger from birds, beasts, and insects, which are only kept off by its resin. But it is difficult to see, for its colour is just that of dead pine needles and its shape is such that it easily slips into crevices. Then the seven or eight small green seed leaves break out of the tough seed coat, and the seedling is now a small tree two inches high. It may have to grow up through grass or bramble, or through bracken, which last is perhaps still more dangerous and difficult. It will probably be placed in a wood or plantation where hundreds of thousands of its cousins are all competing together. "In this case, the struggle for life is intense: each tree seeking for sunlight tries to push its leader-shoots up above the general mass of foliage; but all are growing in height, whilst the lateral branches which are cramped by the neighbouring trees are continually thrown off. The highest branches alone get sufficient light to remain alive, but they cannot spread out freely. They are strictly limited to a definite area; the crown is small and crowded by those of the trees next to it, and the trunk is of extraordinary length."

The above quotation from Albert Fron's *Sylviculture* (Paris, 1903) refers to an artificial forest cultivated and watched over by man. But the trees in such forests have "extra" dangers and difficulties to fight against. Even scientific foresters admit that they are very ignorant of what they are trying to do. In fact, the more scientific they are, the more readily they will confess how little they really know.

Watch a labourer in a nursery transplanting young pine trees; each seedling tree has a long main root which is intended to grow as straight down into the ground as it possibly can. All the other roots branch off sideways, slanting downwards, and make a most perfect though complicated absorbing system. With his large hand the man grasps a tree and lifts it to a shallow groove which he has cut in the soil. Then his very large, heavy-nailed boot comes hard down on the tender root-system. The main root, which ought to point down, points sideways or upwards or in any direction, and the beautifully arranged absorbing system is entirely spoilt. The wretched seedling has to make a whole new system of roots, and in some trees never recovers.

44 All sorts of animals, insects, and funguses are ready to attack our young tree. Squirrels in play will nibble off its leading shoots. Cattle will rub against its bark, and the roedeer, a very beautiful creature, and yet a destructive little fiend from the tree's point of view, nibbles the young shoots and tears the bark with its horns.

A tree's life is full of peril and danger. Yet it is most wonderfully adapted to survive them. Take a knife and cut into the bark of a pine tree, and immediately a drop of resin collects and gathers on the wound. After a short time this will harden and entirely cover the scar. Why?

There are in the woods, especially in Canada and North Russia, hundreds of insects belonging to the most different kinds, which have the habit of laying their eggs in the wood of tree-trunks. In those regions the entire country is in the winter covered with snow and ice for many months. Insects must find it difficult to live, for the ground is frozen to a depth of many feet. Where are the eggs of these insects to be stored up so that they can last through the winter without injury?

There is one insect at least, or rather many, of which the Giant Sawfly may be taken as an example, which have ingeniously solved this problem. She painfully burrows into the trunk of a tree and deposits her eggs with a store of food at the end of the burrow. A drop of resin or turpentine, which would clog her jaws, makes this a difficult task, but, as we find in many other instances, it is not impossible, but only a difficulty to conquer. If it were not for the resin, trees might be much more frequently destroyed by Sawflies than they are.

The larvæ of the Sawfly is a long, fleshy maggot. Just at the end are the strong woodcutting jaws by which it devours the wood and eats its way out as soon as it feels the genial warmth of spring penetrating through the tree-bark. Many other insects hibernate or lay their eggs in tree-trunks.

Some are caterpillars of moths, such as the well-known Goat moth; others are beetles, such as one which burrows between the bark and the wood of apple trees. The mother beetle lays a series of eggs on each side of her own track. Each egg produces a grub which eats its way sideways away from the track of the mother. The track made by these grubs gets gradually wider, because the maggots themselves grow larger and more fat with the distance that they have got from their birthplace. We shall find other instances of burrowing insects when we are dealing with rubber plants.

This resin or turpentine is a very interesting and peculiar substance, or rather series of substances. It is valuable because tar, pitch, rosin, and colophony are obtained by distilling it.

When travelling through the coast forests of pine trees in the Landes of Western France, one notices great bare gashes on the stems leading round and down the trunk to a small tin cup or spout. These trees are being tapped for resin, from which rosin is manufactured. It would be difficult to find any obvious connexion between music and the Giant Sawfly. Yet the rosin used by Paganini and Kubelik has probably been developed in Conifers to keep away sawflies and other enemies. This very district, the Landes in France, was once practically a desert, and famous as such in French history. The soil was so barren that no villages or cultivation were found over the whole length of it. Now that it is planted with trees which are able to yield firewood and rosin, it is comparatively rich and prosperous.

Storms are also very dangerous for tree-life. One can only realize the beauty of a tree by watching a pine or ash in a heavy gale of wind. The swing of the branches, the swaying of the trunk, the balancing support of the roots which, buttress-like, extend out into the soil, give some idea of the extraordinary balance, toughness, and strength in trees. Except in the case of the common umbrella, which is an inefficient instrument in high wind, engineers have never attempted the solution of the problem satisfactorily solved by trees. A factory chimney only feet in height will have a diameter at the base of at least three feet. This means that the height is about seventeen times its diameter. But the Ryeplant, with a diameter at base of 3 millimetres, may be 1500 mm. high! That is, the height is five hundred times its diameter, and the Ryeplant has leaves and grain to support as well as its own stem! In Pine forests on exposed mountain sides there is almost always at least a murmuring sound, which in a storm rises into weird howls and shrieks. With Greek insight and imagination, the ancients supposed that spirits were imprisoned in these suffering, straining pines. That is most beautifully expressed in *The Tempest*, where the dainty spirit Ariel had been painfully confined in a pine tree for a dozen years, and "his groans did make wolves howl and penetrate the breasts of ever-angry bears."

One of the most interesting points in botany depends on the fact that evil conditions of any sort tend to bring about their own remedy. Endymion's spear was of "toughest ash grown on a windy site" (Keats). The prosaic chemical analyses of German botanists have, in fact, confirmed the theory there suggested, for it is found that the wood of trees grown in exposed windy places is really denser and tougher than that of others from sheltered woods.

If one realizes all these dangers from insects, animals, and storms, the height to which some trees grow and the age to which they live become matters for astonishment and surprise.

The tallest trees in the world are probably certain Eucalyptus of Australia, which have obtained a height of 495 feet above the ground.

They are by no means the *longest* plants, for there are certain *rattans* or canes, climbing plants belonging to the Palm family, which may be 900 feet long, although their diameter is not more than two inches. There are also certain Seaweeds in the Southern Ocean, off the coast of Chile, which attain a prodigious length of 600 feet (*Macrocystis pyriferus*, or "Kelp"). That is not so remarkable, for their weight is supported by other plants in the case of the rattans, and as regards the seaweeds, by the water in which they float.

The next in order to the Eucalyptus are those well-known Mammoth or Big trees of California (*Sequoia gigantea*). They grow only in certain valleys in the Sierra Nevada, at an altitude of 5000-8000 feet. Their height is usually given as from 250-400 feet, and the diameter sometimes exceeds thirty-five feet. Since they have become a centre of the tourist-industry in the United

States, various methods have been adopted to make their size more easily realized. Thus a coach with four horses and covered by passengers is (or used to be) driven through a gateway made in one of them. The trunk of another has been cut off some feet from the ground, and a dancing-saloon has been made on the stump. It is at least doubtful if dancing would be very agreeable upon such a cross-grained sort of floor! A complete section of one of them was carried across the United States to make a dining-room table for an American millionaire. The age of one of these trees has been estimated at 3300 years. That is to say that it was a seedling in 1400 B.C., and has been peacefully growing in a Californian valley during all the time when Greece, Rome, Spain, France, Britain, and of course the United States, developed their civilizations. The specimen of the Mammoth tree in the Natural History Museum in London was 1335 years old.

The possible age of many of our common trees is much greater than any one would suppose. The "Jupiter" oak in the forest of Fontainebleau is supposed to be 700 years old. Another oak which was cut down at Bordya, in the Baltic provinces of Russia, was supposed to be about 1000 years old. Other millennial trees are or were another oak and two chestnuts: the oak grew in the Ardennes, the chestnuts still flourish, one at Sancerre (France), and the other the famous specimen on Mount Etna. There are also eight olive trees in the garden of Gethsemane at Jerusalem, which are certainly 1000 years old, and were, according to tradition, in existence in the time of Jesus Christ.

And yet all these trees are mere infants compared to Adanson's Baobab and the Dragon tree of Orotava. The celebrated traveller alluded to visited the Cape Verde islands in 1749 and found inscriptions made by English travellers on the trunk 300 years before his time. From the growth since then, he calculated that some of these trees were about 6000 years of age, and they were 27 feet in diameter.

The record is held by the Dragon tree of Orotava, in the Canary Islands.

When the Spaniards landed in Teneriffe in 1402, its diameter was very nearly 42 feet. It was, however, greatly injured by a storm in 1827, and finally destroyed in 1851. (The wood was then made into walking-sticks and snuffboxes.) The age has been estimated at 10,000 years, or by other authorities at 8000 years only. The "dragon's blood" of the Canaries, a well-known remedy in the Middle Ages, was not, as is popularly supposed, derived from this tree, but was obtained from a totally different plant.

But there is a hazy tradition to the effect that the story of the Dragon which guarded the golden fruit in the island of the Hesperides was nothing but a garbled account of this redoubtable veteran of the plant world.

There is no particular advantage in growing to these enormous heights and clinging to life in this way for hundreds and thousands of years. Nature seems to have found this out and preferred the ordinary pines, oaks, and larches, which are mature in a few hundred years. In a thousand years, ten generations of larch or pine can be produced, and, as each is probably better than its predecessor, a distinct improvement in the type is possible. All these long-lived giants belong in fact to the less highly specialized orders of plants. They are like the primeval animals, the Mammoths, Atlantosauri, and Sabretoothed Tigers.

Yet when we come to think of the many and diverse perils to which trees are exposed, the existence of even these exceptional monsters seems very wonderful.

After a violent storm which had blown down many of the trees in a friend's park, I visited the scene of destruction and discovered what had apparently in almost every instance produced it. Rabbits had overthrown these trees!

They had nibbled away part of the cork and part of the young wood on the projecting buttress-like roots at the base of the tree. In consequence, water, bacteria, and fungus spores had entered at the injured places, and part of the roots had become decayed and rotten. When the gale began to sway them backwards and forwards and a severe strain came on what should have been a sound anchoring or supporting buttress, the rotten part yielded, and these fine, beautiful trees fell a prey to the rabbit.

The influence of forests and timber on the daily life of mankind is a most romantic and interesting chapter in history.

Every savage tribe, every race of man, however degraded or backward, is acquainted with fire. Fuel is therefore a necessity of existence for all savages, and not merely for cooking. There is a very interesting passage in London's *The Call of the Wild*, when the Dog "Buck" in his dreams remembers a hairy man crouching over the fire with Buck's ancestor at his feet, whilst in the darkness all round them the firelight is reflected from eyes of wolves, bears, and even more terrible and dangerous brutes which have now happily vanished from the world. For protection at night fire was an absolute necessity. Even at that long-distant period, therefore, man had commenced to attack the forest. Unless one has had to tend a wood fire for twelve hours, it is difficult to realize what a quantity is required. To prepare fire was a long, laborious, and difficult operation; one piece of wood was placed on the ground and held in position by the toes, a pointed stick was taken between the two palms of the hand and twirled vigorously round and round until the heat was enough to ignite a piece of rotten wood placed as tinder.

Therefore smouldering branches were kept always burning, as they are to-day amongst the Fuegians and some other savages. It was a sacred duty to watch this fire, and the woman (usually old) who was entrusted with the task was very probably put to death if she failed. From this very ancient savage custom probably arose the cult of the Vestal Virgins in Ancient Rome.

Another very important factor in savage life was the canoe or piroque necessary for fishing or to cross lakes and rivers. The first chantey of Rudyard Kipling has a probable theory, and is a beautiful account of how man first thought of using a floating log. They hollowed out the log and "dug out" the canoe, by first lighting a fire on it and then scraping away the cinders; then the sides were pressed out, and it was trimmed and straightened to the right shape. All this was the idea of some paleolithic genius far more persevering and ingenious than any marine architect of our own days.

"Birchbark" canoes are not so common as Dug-outs. The tree, the White or Paper Birch, is found in Canada and the Northern United States; those Indians who discovered that the light, waterproof cork-bark could be fashioned into a canoe made a very great discovery, and indeed it was their canoes that made travel or exploration possible in North America.

When man began to long for a settled permanent home, it was absolutely necessary to find a way of living in safety. Wolves, bears, hyenas and other animals were abundant; neighbours of his own or other tribes were more ferocious and more dangerous than wild beasts. Some neolithic genius imagined an artificial island made of logs in the midst of a lake or inaccessible swamp. Such were the lake dwellings which persisted into historic times, and which are indeed still in existence in some parts of the earth.

The trees were abundant; they could be felled by the help of fire and an axe, and the lake dwelling gave a secure defence. The wood of some of the piles supporting the great villages in Switzerland seems to be still sound, though it has been under water for many centuries. Some villages are said to have required hundreds of thousands of trees.

The forest afforded man almost everything that he used, bows and arrows, shelter, fuel, and even part of his food.

Nuts and fruits would be collected and when possible stored. In seasons of famine, they used even to eat the delicate inside portion of the bark of trees.

But as soon as the first half-civilized men began to keep cattle, sheep, and especially goats, more serious inroads still were made upon the forest. Where such animals are allowed to graze there is no chance for wood to grow (at any rate in a temperate country). The growing trees and the branches of older ones are nibbled away whilst they are young and tender. The days of the forest were nearly over when cultivation commenced. Dr. Henry describes the process of "nomadic" culture in China as follows: "They burn down areas of the forest; gather one or two crops of millet or upland rice from the rich forest soil; and then pass on to another district where they repeat the destruction." A very similar process of agriculture existed until the eighteenth century in Scotland.

The logs in the long train of rafts are of bamboo tied together. In spite of their fragile nature the lumbermen are so fearless and agile that they cleverly steer the frail bundles with but few accidents.

Thus the forest was being burnt or cleared for cultivation. It was devastated by black cattle, goats, and other animals, and it was regularly exploited for fuel and building every day by every family for centuries.

It is not, therefore, surprising that the ancient forests in Britain have disappeared. Dr. Henry mentions one square mile of virgin forest on the Clonbrock estate in Ireland. The *Silva Caledonica* of the Romans is said to exist in Scotland at the Blackwood of Rothiemurchus, at Achnacarry, and in a few other places. Of the original oak forest, which covered most of England and Southern Scotland, not a vestige (so far as is known to the writer) remains to-day.

There are in places very ancient forests. A few miles from Retford are considerable remains of Sherwood Forest, which is for ever associated with that genial bandit Robin Hood. One huge oak (called the Major) has or used to have a keeper always on guard and paid by Lord Manvers, but there are hundreds of aged oaks all round it. Then there is the Knightwood Oak and some other ancients in the New Forest.

But it is not certain that these even date so far back as the time of Canute, for so far as the New Forest is concerned, it seems that this was formed either by Canute or by William I. The Saxons seem to have destroyed most of the English forests.

In Scotland oak forest existed as far north as the Island of Lewis, in Caithness, Dornoch, Cromarty, and along Loch Ness, as well as in every county south of these. The deer forests and grouse moors, now desolate, whaup-haunted muir-land and peat mosses, were flourishing woods of magnificent Scots fir at no very distant period. They ascended the hills on the Cairngorms to 1400 or 1500 feet, and in Yorkshire to 2400 feet.

Even in remote historical times, such as those of Canute, the forests had become seriously and dangerously destroyed. This king was apparently the first to artificially protect the woods as a hunting preserve. He was followed by William the Conqueror and other sovereigns. The game preserves of the landed proprietors to-day are, of course, the remains of the same custom.

Fortunately, however, we do not kill poachers or cut off their right hands, and we do not cut off the forepaws of poaching dogs, as used to be done in medieval days.

This connexion of forests with game no doubt prevented the entire disappearance of wood, but when, as is the case in England, the comfort of pheasants is thought of more importance than the scientific cultivation of forests, the result is often very unfortunate.

The use and value of timber is, however, too important a matter to take up at the end of a chapter.

CHAPTER IV. ON FORESTS.

THE FORESTS OF THE COAL AGE—MONKEY-PUZZLE AND GINKGO—WOOD, ITS USES, COLOUR, AND SMELL—LASTING PROPERTIES OF WOOD—JARRAH AND DEODAR—TEAK—USES OF BIRCH—NORWEGIAN BARQUES—DESTRUCTION OF WOOD IN AMERICA—PAPER FROM WOOD PULP—FOREST FIRES—ARID LANDS ONCE FERTILE—BRITAIN TO BE AGAIN COVERED BY FORESTS—VANISHED COUNTRY HOMES—ASHES AT FARMHOUSES—YEWS IN CHURCHYARDS—HISTORY OF MAN VERSUS WOODS IN BRITAIN.

WHAT was the first tree like? That is a very difficult question to answer. Perhaps the first forests were those of the great coal period, of which the remains, buried for untold ages in the earth, became the coal which we now burn.

The flames and red-glowing heat of a fire are the work of the sunlight which fell in these long-past ages through a steamy, misty atmosphere, upon these weird, grotesque vegetables, unlike anything which now exists upon the earth. Their nearest allies amongst living plants are the little club-mosses which creep over the peat and through the heather in alpine districts.

Of course no one can say exactly what these coal forests were like. But although some modern authorities have questioned the general accuracy of the descriptions of Heer and others, yet, as they have not given anything better in the way of description, we shall endeavour to describe them according to our own beliefs, and as they probably existed in the Lanarkshire coalfield and other places in Britain.

In that gloomy mirk of the Carboniferous epoch, an observer (if there had been any) would have dimly perceived huge trunks rising to sixty or eighty feet and divided at the top into a very few branches. All branches were covered over by comparatively quite small leaves. Not a bad idea of the Sigillarias, Lepidodendrons, etc., which made the forest and can be obtained by carefully looking at a pan of Selaginella such as one finds in almost every botanical garden, and imagining this to be eighty feet high. Through the bottomless oozy slime which formed the ground, horizontal runners and roots penetrated in every direction. Great fern-like plants might be observed here and there. Sluggish rivers meandered slowly through these forests, carrying silt and refuse (their deposits are our Cannel coals). In the water and in pools, or perhaps in the mud, were curious waterferns with coiled-up crozier-like leaves. Perhaps horsetail-like plants of huge size might have formed great reed-beds to which those of to-day are as a plantation of one-year-old firs is to a pine forest that has lasted for a century.

Fishes and crustaceans, or lobster-like creatures, crawled and squattered through the slime, pursued by salamander-like animals with weak limbs and a long tail. Some of these latter were seven to eight feet long. Millipedes, scorpions, beetles and maybugs existed, and huge dragonflies preyed on them.

But there is one very ancient group of trees, the Araucarias or Monkey-puzzles, which are by no means uncommon even now. The ordinary one (*Araucaria imbricata*) is often planted in the British Isles, and it has, if you look closely at it, a most peculiar appearance. It is like the sort of tree that a child would draw; it is a clumsy attempt at one, and very different from the exquisite irregularity of the ash or oak.

Its leaves are especially curious: they cover the branches very closely, and are hard, rigid, and spiny. Its cones, though of the nature of pine-cones, are yet quite unique. The seeds are edible, and used to be an important article of diet to the Indians on the slopes of the Chilian Andes, where monkey-puzzle forests used to exist. This of course is a very out-of-the-way region; other species of Araucaria are found scattered about the world in a most perplexing manner. One kind grows in Norfolk Island, in the Pacific; another occurs in the inner mountainous districts of Brazil; there are some in Australia and others in New Caledonia.

But in the Jurassic period of geology, in the age of ammonites and gigantic lizards and crocodiles, Araucarias were the regular, ordinary trees. They grew all over Europe, and apparently as far north as Greenland, and, indeed, seem to have existed everywhere.

Perhaps the spiny leaves discouraged some huge lizard, perhaps Atlantosaurus himself (he was thirty feet high and one hundred feet long), from browsing on its branches. Perhaps the Pterodactyls, those extraordinary bird or bat-like lizards, used to feed upon the seeds of the monkey-puzzle, and carried them in their toothed jaws to New Caledonia, Australia, and Norfolk Island. Other improved types have driven the monkey-puzzles from Europe, Asia, and Africa, and taken their places, but in out-of-the-way districts of South America and Australia they are still able to hold their own.

An ally of theirs, the Ginkgo or Maidenhair tree, seems to have been extremely common in certain geological periods. To-day it has almost entirely disappeared. A few trees were discovered in certain Chinese temples, where they had been preserved as curiosities for centuries, but it is almost extinct as a wild plant. The Bigtree group was a companion of the Ginkgo in its flourishing period. So also were the Sago palms or Cycads. All the ordinary trees, Pines, Oaks, Beeches, and the like, did not appear upon the earth's surface till a much later period.

The most important economic product of trees is the timber which they furnish. Wood, as we have tried to show in the last chapter, has been always of the greatest importance to mankind. It is easily worked, durable, buoyant, and light, and it is used for all sorts of purposes.

Silver fir, which is accustomed, when growing, to be continually swayed and balanced by the wind, is preferred for the sounding-board of pianos and for the flat part of violins, whilst Sycamore or hard Maple is employed for the back and sides of the latter.

But there are enormous differences in different kinds of woods. The colour of wood varies from white (Beech), yellow (Satinwood), lemon-yellow and bluish red (sap and heartwood of Barberry), to dark and light brown mottled (Olive), black (Persimmon), and dark brown (Walnut). Some woods have a distinct smell or perfume. Cedarwood, Sandalwood, Deal, and Teak, are all distinctly fragrant. The Stinkwood of South Africa and the Til of Madeira have an unpleasant smell.

More important in practice are the differences in the hardness and weight of wood. The Ironwood of India cannot be worked, as its hardness blunts every tool. It requires a pressure of something like 16,000 lb. to force a square-inch punch to a depth of one-twentieth of an inch in *Lignum vitæ*. Even Hickory and Oak (if of good quality) require a pressure of 3200 lb. to the square inch to do this. On the other hand the Cotton tree of India (*Bombax malabaricum*) has exceedingly soft wood. It is quite easy to drive a pin into the wood with the fingers.

Some woods are far too heavy to float: many tropical woods are especially very weighty. Perhaps the Black Ironwood, of which a cubic foot weighs 85 lb., is the heaviest of all. But the same volume of Poplar, Willow, or Spruce does not weigh more than 24 lb.

There are many ancient and modern instances of the extraordinary way in which timber lasts when at all carefully looked after. Thus the Cedar which "Hiram rafted down" to make the temple of Solomon (probably Cedar of Lebanon) seems to have been extraordinarily durable. Pliny says that the beams of the temple of Apollo at Utica were sound 1200 years after they were erected.

Cypress wood (*Cupressus sempervirens*) was often used to make chests for clothes because the clothes moth cannot penetrate it, and it also lasts a very long time. There is a chest of this wood in the South Kensington Museum which is 600-700 years old. The Cypresswood gates of Constantinople were eleven centuries old when they were destroyed by the Turks in 1453. The fleet of Alexander the Great, and the bridge over the Euphrates built by Semiramis, were made of Cypress. This wood seems to have been of extraordinary value to the ancients, and was used for mummy cases in Egypt, for coffins by the Popes, as well as for harps and organ pipes.

Perhaps the most valuable woods are Box, which is used for woodcuts, and Walnut, which used to be highly prized for gun-stocks, as much as £600 having been paid for a single tree.

But the most interesting histories of trade in timber belong to the commoner and more usual woods. The great woods of Jarrah (*Eucalyptus marginata*) cover 14,000 square miles of Australia, but they are being rapidly cut down and sawn up into small blocks to be carried right across the world in order to form the pavement which London cabmen and cab-horses prefer to any other.

One remembers also the beautiful Deodar forests of Afghanistan, and the Himalayas. Logs of deodar were floated down the rivers to form bridges or temple pillars in Srinagar, the capital of far Cashmere. Nowadays great "slides" are made, winding down into the valleys from the recesses of the hills. When winter approaches, water is sprinkled on the logs which make the slide; this freezes and forms a slippery descending surface, down which the deodar timber rushes till it reaches the low ground, where it is cut up into railway sleepers and takes part in the civilizing of India.

The fragrant Teak has an oleoresin which prevents the destructive white ants from attacking it; it is the most valuable timber for shipbuilding, and grows in many places of India, Malaysia, Java, and Sumatra. It floats down the rivers of Burmah, coming from the most remote hill jungles, and elephants are commonly used at the ports to gather the trunks from the water and pile them ready for shipment.

The Birch is carried all the way from Russia to Assam and Ceylon, in order to make the chests in which tea is sent to England and Russia (native Indian woods are also used). It is also used in the distillation of Scotch whisky, for smoking herrings and hams, for clogs, baskets, tanning, dyeing, cordage, and even for making bread.

But one of the most curious and interesting sights in any seaport is sure to be an old white Norwegian or Swedish sailing barque or brigantine. She will have a battered, storm-beaten appearance, and is yet obviously a comfortable home. The windows of the deck-house may be picked out with a lurid green. The tall, slowmoving, white-bearded skipper and his wife, children, and crew, not to speak of a dog and cats, have their home on this veteran "windjammer." She carries them from some unpronounceable, never-heard-of port in Norway, all over the world. You may see her discharging a cargo of deal plank, through the clumsy square holes in her stern, in a forgotten Fifeshire village, in Madagascar, in China, or in the Straits of Magellan. All her life she is engaged in this work, and her life is an exceedingly long one, to judge from the Viking lines on which she is built.

Moreover, her work is done so economically that it used to be much cheaper to use her cargo in Capetown than to utilize the beautiful forests of the Knysna and King Williamstown.

But there are not wanting signs that the forests of Norway, of Sweden, and even those of the United States, are doomed.

It is said that seven acres of primeval forest are cut down to supply the wood which is used up in making the paper required for one day's issue of a certain New York journal. What a responsibility and a source of legitimate pride this must be to the journalists! Let us hope that the end justifies the means.

Boulger calculates that in 1884 all the available timber from 4,131,520 acres of Californian Redwood was used in making the sleepers of the railways then existing in the United States.

He finds that no less than 18,000,000 acres of forest are necessary to keep up the supply of sleepers for the old lines and to build new ones.

So that, if we remember the wood required for paper, firewood, and the thousand other important requisites of civilized man, the United States must soon exhaust her supply and import wood.

Then will come the opportunity of British North America. The Southern forest of Canada, which extended for 2000 miles from the Atlantic to the head of the St. Lawrence, has indeed gone or is disappearing into pulpwood and timber, but there is still the great Northern forest from the Straits of Belleisle to Alaska (4000 miles long and 700 miles broad), and in addition the beautiful forests of Douglas Spruce and other trees in British Columbia covering 285,000 square miles.

It is the wood-pulp industry which is at present destroying the Canadian forests. The penny and halfpenny papers, and indeed most books nowadays, are made of paper produced by disintegrating wood: it is cheap, and can be produced in huge quantities; nevertheless it is disquieting to reflect that probably nineteen-twentieths of the literary output of the twentieth century will be dust and ashes just about the same time (some fifty years) that the writers who produced it reach the same state.

Yet, considering the amount daily produced to-day, the future readers of fifty years hence who are now in their cradles, may consider this a merciful dispensation of Providence.

One very curious use of wood may be mentioned here. Near Assouan, on the First Cataract of the Nile, one discovers broken granite or syenite needles, which had been intended by the ancient Egyptians for monuments. Where the broken pillar lies, there are rows of wedge-shaped holes cut in the rock.

They used to drive in wedges of dry wood and then wet them with water. The expansion of the wood split the rock, though this is hard granite or syenite. Very often the process failed because the stone cracked. The same method is said to be still used in some quarries.

The destruction of the forest is really necessary. Most of the corn land and rich pasture of the world has been at one time forest. It could scarcely be such fertile soil if it had not been for the many years during which leaf-mould fell on it, and the roots broke up and penetrated the subsoil below. Canada, Russia, and the United States are now passing through the same experience as that of Great Britain in the time of the Romans, Saxons, and Danes.

But there is terrible waste by fire.

When the trees become dry and withered in the height of summer in either India or the United States, some careless tramp may throw aside a lighted match. If a fire once starts, it spreads with enormous rapidity; great clouds of smoke roll over the surrounding country, and every village sounds the alarm. Everybody rushes to help and try to stop the conflagration, or if too late hurriedly saves whatever he can get of his possessions. His log hut and all the accumulations of years of saving may be turned into a heap of ashes in a very few minutes.

But the crackling of the leaves and the flaming twigs and scorching bark make such a volume of fire that nothing which man can do is of any avail.

Of course every beast, every bird and insect is in the greatest possible danger.

This is how a fire in New Zealand has been described by Mr. William Satchell:—

"For a while it seemed that the battle must go to the wind, the fiery monster withdrew, lay hidden, roaring angrily in the dry heart of the woods; then insidiously he stretched forth his glittering arms, first one, then another, and locking the shuddering trees in an irresistible embrace, sprang once again erect. In an instant the whole bush from edge to edge became a seething, rocking mass of flames.

"'Fire! Fire!'

"Then, insignificant no longer, transfigured rather beyond all living possibilities of loveliness, the bush stood revealed to its centre. It became less a fire than an incandescence, waxing in brilliance to the point when, as it seemed, it must perforce burst into indistinguishable flame. Every leaf and twig of that fairy forest was wrought and hammered in virgin gold, every branch and trunk was a carved miracle of burnished copper. And from the golden leaves to the golden floor, floatingly or swiftly, there fell an unceasing rain of crimson flame petals, gorgeous flame fruits. Depth after depth stood revealed, each transcending the last in loveliness. And as the eye sought to penetrate those magic interiors there seemed to open out yet farther vistas, beyond belief beautiful, as of the streets of a city incorruptible, walled and towered, lost in the light of a golden incomparable star."

"'Fire! Fire!'

"In the face of that vision of glory the cry rang out with all the ineptitude and inappropriateness of the human weakling. On one side the titanic forces of nature, inexorable, eternal; on the other the man, frail of body, the creature of an hour, matching himself against them.

"'Fire! Fire!'

"Sheltering his face from the insufferable heat, the Swede hammered madly at the solid house-door. At the back, now utterly unapproachable, the kitchen, the roof, and a part of the main wall were already in flames. A few minutes—five at the most—would complete the demolition of the house. To right and left the great trees one after another went off like rockets, the roar of their burning foliage shaking the very earth. A deafening crashing of falling timber came at intervals from the bush beyond."

In some countries the destruction of the forests has had a very serious effect on the climate. The rain which falls upon a forest is partly absorbed by the leaves, and but a very small part of it is carried off by burns and streams: most sinks down into the forest soil, and is only gradually given back again after being taken in by the tree roots and evaporated by the leaves.

But bare hills denuded of wood allow most of their rain to rush down to the sea in dangerous spates of the rivers and burns, and then the ground becomes afterwards very dry and burnt up. There are very many countries now barren and desolate because they have been robbed of the beautiful forests which once covered the springheads and mountain valleys.

Perhaps Palestine is one of the worst instances. But it is when we remember Babylon, Nineveh, and all the cities of the coast of Asia Minor, as they were even a thousand years ago, and compare their present barren, desolate condition, that the full meaning of mountain forests becomes clear.

Where once there were thriving, prosperous cities with enormous populations, now the goats graze or a few miserable peasants carefully husband the water of a few miserable streams. The same thing has happened in Mauritius, in the Cape Verde and Canary Islands, and in many other places.

But men are now beginning to see how dangerous the destruction of forests may be, and in many countries and especially in Britain, new forests are being planted. Perhaps in time we may grow in Britain so much timber that we shall gain something like £32,000,000 a year, which is what we spend on imported woods.

At present plover, whaups, snipe, and grouse, or useless red deer, inhabit what was once the Caledonian forest, and every thousand acres of such land nowadays supports perhaps one shepherd and half a gamekeeper. But when it is planted again with woodlands it will afford a living to at least ten foresters, and surely a whole gamekeeper as well.

In the lowlands of Scotland and in England one often discovers, in walking over the hills, remains of cottages and farmhouses which have now vanished. The people have gone into the towns, and the healthy yeomen and farmers' boys have become weak-chested factory hands and hooligans. Such sites of old farms can often be recognized by a patch of nettles, and especially by eight or nine ash trees. These were always planted near the houses to give a ready supply of wood for spears. The ash, "for nothing ill," as Spenser puts it, would be available also for repairing the handles of tools, carts, etc. Some authorities say that it was the law of Scotland that these eight or nine ash trees should be planted at every "farmtoon."

So also, when forests began to vanish in England, laws were made to the effect that yew trees should be planted in every village churchyard. Probably this was to ensure a good supply of bows for the English archers, who, like the Scottish spears, were the best soldiers of their kind in Europe.

Such fires frequently occur in New Zealand, and the Maoris have to fly for their lives.

So that if we try to compare the conditions of man and of the forests in Great Britain from the earliest days, it would be something like this: —

1. When the earliest inhabitants lived on shell-fish, seabirds' eggs, nuts, and fruits, almost the whole country was covered by oak, Scotch fir, or birch forests.

2. When man was a hunter of reindeer and other deer, horses, cattle, and birds, he used much wood for fires and for building his lake dwellings.

3. When man kept herds of swine to eat acorns, black cattle, goats, and ponies, there would be many clearings and a great deal of open wood in which the cattle roamed about.

4. When man grew corn and other plants, the forest vanished altogether. Dr. Johnson said he scarcely saw a tree between Carlisle and Edinburgh. Yet first the King, then the Barons, had their parks and woodlands for preserving game. Moreover, the yews in the churchyards of England, and the ash trees by the Scotch farmtoons and peel-towers, were carefully looked after.

5. When great towns arose, and men became factory hands and steel workers, rich men began to make plantations in the lowlands, and to use the depopulated highlands for grouse moors and deer forests.

6. When men become wiser than they are now, it will be seen that great forests are necessary on all waste-land and barren places, both to keep a healthy country population and because it will pay.

CHAPTER V. FLOWERS.

MAN'S IDEAS OF THE USE OF FLOWERS—SPRENGEL'S GREAT DISCOVERY—INSECTS, NOT MAN, CONSULTED—POLLEN CARRIED TO SET SEED—FLOWERS AND INSECTS OF THE WHINSTONE AGE—COAL AGE FLOWERS—MONKEY-PUZZLE TIMES—CHALK FLOWERS—WIND-BLOWN POLLEN—EXTRAVAGANT EXPENDITURE OF POLLEN IN THEM—FLOWER OF THE PINE—EXPLODING FLOWERS—BRILLIANT ALPINES—INTENSE LIFE IN FLOWERS—COLOUR CONTRASTS—LOST BEES—EVENING FLOWERS—HUMMING BIRDS AND SUNBIRDS—KANGAROO—FLORAL CLOCKS—AGES OF FLOWERS—HOW TO GET FLOWERS ALL THE YEAR ROUND—INGENIOUS CONTRIVANCES—YUCCA AND FIG—HORRIBLE-SMELLING FLOWERS—ARTISTIC TASTES OF BIRDS, INSECTS, AND MAN.

FOR many centuries flowers were considered as pleasing and attractive decorations stuck about the world in the same way as they are put in a drawing-room in order to give people pleasure. Very soon they were found to be extremely useful in poetry, sometimes to point a moral or disguise a sermon, like the primrose in *Peter Bell*, but more generally to produce a good impression on the BELOVED OBJECT. Burns puts the usual view of flowers very nicely in the following: "But I will down yon river rove amang the woods sae green, and a' to pu' a posie to my ain dear May." Possibly this is the meaning also in the exquisite lines of Shakespeare about the pansy:—

"Yet marked I where the bolt of Cupid fell:
It fell upon a little western flower,—
Before milk-white, now purple with love's wound,—
And maidens call it, love-in-idleness."

Even if there is no particular meaning, the "little western flower" gives point and beauty to the lines.

People only began to understand flowers about the year 1793, when Christian Conrad Sprengel, Rector of Spandau, near Berlin, published a very interesting work. He had discovered that the beauty of flowers and their colour and shape were by no means intended solely to please *human* eyes, but that they were designed to attract and allure the eyes of *insects*. Before his time there had been many guesses. Indeed, Theophrastus (born 371 B.C., and often mentioned in this work) seems to have quite well understood why flowers produce pollen, and that the fruit would not set and form seed unless pollen was carried to the female part of the flower. He mentions that the Pistacio has both male and female plants, and that Palms only form dates when the pollen is carried to the female tree. This experiment with the Date-palm was tried in 1592 by an Italian (Alpino) in an Egyptian tour, and the Englishman, Jacob Bobart, the Pole, Adam Zaluzianski (the latter in the same year) confirmed the general idea. Then in the year 1694 Rudolp Jacob Camerarius, a German, carried on a few more experiments, but no real definite advance was made until 1793, in the very midst of the French Revolution.

The great point of Sprengel's discovery was in its being an intelligible explanation of the reason why flowers have bright colours, scent, and honey. At his time and indeed for many years afterwards, botanists looked on the stamens, petals, and other parts of the flower exactly in the way that a stamp collector looks at punctures and postmarks, that is without thinking about their meaning. Now we find that they are always designed to fulfil a perfectly definite purpose, and that all their details are contrived accordingly.

This purpose is to carry the pollen from the stamens of one flower to the stigma of another. The pollen can usually be recognized as a yellowish or reddish dust formed in the stamens; this dust is generally rubbed off on an insect's proboscis or on part of its body. When the insect reaches another flower the pollen is scraped off by a sticky or gummy stigmatic surface. When the pollen has been placed on this surface it grows, germinates, and part of it unites with the egg-cell of the young seed.

The latter is then, and not till then, able to become ripe and mature. It may be compared to cross-breeding in animals, though the process does not exactly correspond.

But all flowers do not require insects to carry their pollen. In early geological periods we do not find any flowers like those that now exist, nor in those early times were there any flies, bees, or butterflies.

The cockroach seems to have existed in Silurian (whinstone) times, and many gigantic and extraordinary insects lived in those damp forests of ferns, club-moss, and horsetails, of which the remains now form our British coalfields. Mayflies, plantbugs, and especially dragonflies (some of them with wings two feet across) existed, but none of these insects are of much use as pollen-carriers.

Even much later on, when screw pines, monkey-puzzle trees, ginkgos, and bamboos formed the forests and woods of Europe, crickets and earwigs existed; but it is not until that geological period in which the chalk was formed (the Cretaceous age) that fossil plants like most of those now familiar to us occur. These had flowers intended for insects, and with the fossil plants we find the fossils of the insects that visited them. Bees, butterflies, and ordinary flies appeared upon the scene just as soon as there were flowers ready for them. Mr. Scudder has even found the fossils of certain plants, and with them the fossils of butterflies closely allied to the present butterflies which now live on present trees allied to those fossils!

How then was the pollen of the first flowers carried?

It was in all probability blown by the wind or carried in water. Even now poplars, alders, birches, and oaks rely chiefly upon the wind to carry their pollen. These plants were amongst the first of our modern flora to appear upon the earth. Some of them possess very neat contrivances suited to the wind. The catkins of the alder, for example, hang downwards, so that each little male flower is protected from rain by a little scale or bract above it. The pollen is very light, dusty, or powdery, so as to fly a long distance. The Scotch fir (*Pinus sylvestris*) has male flowers in little cones. These are upright, and the pollen of each stamen drops on to a small hollow on the top of the stamen below. It is then blown away by the wind on a fine dry day, but it is not allowed to get out in wet weather. It is said that vast clouds of pine pollen occur in America, and that the water of certain lakes becomes quite yellow and discoloured by it at certain seasons. Each little particle of pollen has two minute caps or air-balloons which give it buoyancy, so that it can float easily immense distances.

A curious little herb, the Wall Pellitory, and another foreign species, the Artillery plant, produces small explosions of pollen. When it is touched, there is a little puff or cloud of dusty pollen. Even the common Nettle does the same on fine dry days when it is in full flower.

But of course this carrying of pollen by the wind is a very expensive arrangement. It is so much a matter of pure chance that a grain arrives at its right destination. Suppose that a flower is giving out clouds of pollen, then the chance of a pollen grain reaching a female flower only five feet away is very small, even if the stigma of the female flower is a quarter of an inch in diameter. The chance of pollen reaching it will only be about 1 to 1440; 1439 pollen grains will be wasted for every one that reaches the stigma. But even this is not quite a fair calculation, for if the female flower is not down wind, none will reach it at all!

But if an insect goes to the catkin of an alder or any other male flower, it will see the red points of the stigma and will very likely go there at once. This shows how much more reasonable and efficient insects will be.

The immense majority of flowers are, in fact, purple, blue, red, yellow, or white, so that they are conspicuous, and stand clearly out against the green of their leaves. It is well known to all who have arranged flowers for the table that the green of the leaves of different plants varies greatly in its shade and tint. Many greens do not match special flowers at all, but it is the fact that the green of any one plant is always quite harmonious, and agrees well with its own flowers!

Besides varied and beautiful colours, sweet or strong scents and supplies of honey or nectar are provided for insects.

How did flowers manage to produce all these attractions? No one has answered that question. We know in a general sort of way that the parts of flowers are modified leaves, and that petals and stamens become yellowish or pure white because they do not form green colouring matter like ordinary leaves.

It is also known that on the Alps or on any high mountain, where the air is pure and the sun strong, flowers become rich, brilliant, and vivid. In such places as the "Jardin" near Mont Blanc, the pure, deep, rich blue of gentians, the crimsons, reds, and purples of other flowers, impress the most casual and unobservant traveller. "White and red, yellow and blue, brown and green stand side by side on a hand's breadth of space." In that strong mountain air, also, perfumes are stronger, purer, and of finer quality than in the lowlands. There is a more intense, active, and vigorous life going on in flowers than is required by the more prosaic industries in other parts of a plant. Flowers also often live at a higher temperature than the surrounding air.

Kerner has described how the little flowers of Soldanella penetrate the snow by actually melting a passage for themselves through it .

This high temperature and vigorous life, shown also by the rapid transpiration of flowers, seems to hint that colours and perfumes appear in consequence of rapid chemical transformations.

It was, of course, by degrees that the extraordinary variation in colour, which exists in nature, came about. No doubt bees, bumble-bees, wasps, and the more intelligent flies were improved and developed æsthetically. We can almost tell by looking at a flower what sort of insect probably visits it.

Not only so, but there are the neatest imaginable contrasts and blends of colour. The common Bluebeard Salvia, e.g., has the uppermost leaves (three-quarters to an inch long) of a deep, rich, blue-purple, which the roving Bumble-bee will see from a long way off. The Bumble-bee flies to this great splash of her favourite hue and for a second buzzes angrily, then she notes the small *bright-blue* patches on the upper lips of the small flowers below the leaves which are set off by *white* hairs of the upper and *yellow* hairs of the lower lip.

That bees really do understand and are guided by colour may be gathered from the following unfortunate accident. A certain hive of bees which had been brought up in a blue-striped skep became accidentally scattered. They tried to find their way back to their old home, but many strayed, and it was noticed that they had tried to enter the doors of every blue hive, which were strewn with the bodies of the unfortunate intruders.

The rich blue-purple of Aconite, the dark strong red of the Woundwort (*Stachys silvatica*) are specially beloved by bumble-bees and hive-bees. Butterflies like any bright colour. Those flies which have a long, sucking proboscis, resemble the bees in their tastes, but all these insects are quite capable of finding out where they can get honey most easily, and visit flowers whatever the colour may be.

A very strange and wonderful fact is that quite a number of plants prefer the dark, or rather the dim, mysterious light of the gloaming. Then the Honeysuckle, the Evening Campion, the Night-scented Stock, Tobacco, and Schizopetalon give out their strongest scent and open out their white flowers as widely as possible. That is because they wish to attract the owlet moth and others which come out at this time, when there are fewer enemies and more security. If you look at any of these moth-flowers at mid-day, they are for the most part closed up, they are not particularly attractive, and they are giving out very little scent. The contrast to their condition in the evening is most striking.

Not only insects but birds are used to carry pollen. The gorgeous little humming birds, with their brilliant metallic crimson, bronze-green, and purple, are of the greatest importance in the New World. In the Old World they are replaced by the tiny *Nectarinidæ* or Sunbirds, with breastplates almost as exquisitely jewelled. They prefer the most gorgeous reds and scarlets, such as that of *Salvia horminum*, *Lobelia cardinalis*, and the like. Fuchsias are regularly visited by them in Tierra del Fuego, where sometimes they may be seen busily at work during a shower of snow. In South Africa they seize the stem of a Redhot Poker (*Tritoma*) (*Kniphofia macowanii*), and twisting their little heads round, they suck the honey from every blossom in succession. Still more interesting it is to see them perched on the edge of one of those great tumbler-like heads of Protea (e.g. *P. incompta*) and dipping their slender curved beaks repeatedly into the flowers. Then the little male bird will alight on a branch and make the most elaborate preparation for a song of triumph. Although helped out by fluttering of wings and much display of feathers and tail, the song is a very faint cheep of the feeblest description, and very difficult to hear.

Not only birds but even animals are sometimes called into the service. There is a group of small mammals which live on the honey of flowers. Even the Kangaroo is said to occasionally take a draught of nectar from some of the cup-like flowers of the Australian Dryandra (*Proteaceae*).

But one of the most interesting and extraordinary facts is the manner in which flowers fit in. They begin early in the morning: one blossom opens out and then another; all endeavouring to catch the attention of some passing insect. *Allionia violacea* opens at three or four a.m., and closes about eleven or twelve. Some wild Roses open about four or five in the morning, as well as the Chicory, Roemeria, etc. Virginian Spiderwort, Dandelion, and Nightshade are ready at six in the morning. A great many (Buttercups, White Water Lily, etc.) are open by seven a.m. Most of these early flowers are shut at noon. Others begin to close about three or four in the afternoon. The regular evening moth-flowers open about six p.m., though *Cactus grandiflorus* does not open till nine or ten p.m., and closes at midnight. Extraordinary as these variations seem, they are easily explained. Some open early because there are then few competitors. By far the greater number are open from nine a.m. till one or two p.m., because those hours are the favourite working time of most insects.

Flowers live for very different periods. That of the Wheat only lasts for fifteen or twenty minutes (its pollen is carried by wind), and is then over. There are others, Hibiscus and Calandrinia, which only remain open for three or four hours, but a Foxglove will last six days, a Cyclamen ten days, whilst Orchids may last for from thirty to eighty days (*Cypripedium villosum*, seventy days, *Odontoglossum Rossii*, eighty days).

Thus the sun every day through the summer, as he calls into life new swarms of insects, sees at every hour of the day new flowers opening their petals to his genial warmth and ready for the new bees and flies. The development of the flower and that of its insect are probably simultaneous, and equally regulated by the sun's warmth. Moreover the opening periods do not merely fit in during the day, but each flower has its own special month, and even in Scotland there is no month in which some flower may not be found in bloom. Any stray wandering insect can get its draught of honey at any season of the year.

This is a matter of some importance for those who keep bees, and the following list may be of some use. *February*: *Crocus vernus*, Snowdrop, Black Hellebore, and Hazel. *March*: The preceding, *Arabis alpina*, Bulbocodium, *Cornus mascula*, *Helleborus fœtidus*, Giant Coltsfoot, Gooseberry, various species of Prunus and Pyrus, Willow. *April*: The preceding as well as *Adonis vernalis, Barbarea vulgaris, Brassica napus.*

It is not worth while noting those that bloom from May to September, for there are hundreds of good bee-flowers in these months. In *October*: Borage, Echium, Sunflowers, *Lycium europæum, Malope grandiflora,* Catmint, Tobacco, Ocimum, Origanum, *Phacelia tanacetifolia,* and others. Most of these last into November. In December and January very few plants are in bloom. The following have been noted at Edinburgh Botanical Gardens: *Dondia epipactis, Tussilago fragrans,* Snowdrop, *Geum aureum,* Hepatica, *Primula acaulis, P. veris, Aubrietia deltoidea, Crocus imperati, C. suaveolens, Erica herbacea alba,* Helleborus (3 species), *Polygala chamaebuxus, Andromeda floribunda;* also Sir H. Maxwell mentions *Azara integrifolia, Hamamelis arborea,* and *Chimonanthus fragrans.* Of wild plants, Chickweed, Whin or Furze, *Lamium purpureum,* and Dandelion can generally be found in the depth of winter.

The contrivances which can be found in flowers, and by which the insect is forced to enter exactly along the proper path, are endless. Each flower has some little peculiarity of its own which can only be understood by thoroughly examining the plant itself. It is not therefore possible to do justice to the ingenuity of flowers in a work of this sort. There are orchids which throw their insect visitors into a bath of water, so that they have to crawl with wet wings up a certain path where they touch the pollen masses and stigma; others which hurl their pollen masses at the visitor. In the Asclepiads a groove is provided into which the leg of the insect slips, so that it has to struggle to get its foot out, and must carry off the pollen masses, though it often fails and leaves its leg behind. Some Arums and Aristolochias have large traps in which they imprison the insects, and only let them go when they are sure to be pollen-dusted. In one of these flowers there are transparent spots on the large petal-prison, which so attract the insects that they remain opposite them instead of flying out (just as flies do on a window-pane). Salvia has a stamen which is like a see-saw on a support; the bee has to lift up one end, which brings the other with its pollen flat down on to its back. The Barberry has a sensitive spot on its stamen; when the insect touches the spot, the stamen springs up suddenly and showers pollen upon it. In Mimulus the two flaps of the stigma close up as soon as they are touched, which will be when they have scraped off any pollen; then when the creature withdraws, covered with the flower's own pollen, none of this can be left on its own stigma, as this is shut up. But instead of reading, one should watch a bumble-bee visiting the Foxglove flowers. The sight of her busily thrusting her great hairy body into the bell, which almost exactly fits her shape, while she gurgles with satisfaction, will teach the reader far more about the romance of flowers than many pages of description. If he then carefully examines the flower, he will see how the honey, the arched converging stamens, and the style, are placed exactly in the right place and where they will have the most effect.

One orchid, *Angraecum sesquipedale*, has a spur eighteen inches long, and the great Darwin suggested that there must be an insect somewhere with a tube long enough to reach the honey. Such an insect, a large moth, was actually brought home from Madagascar, the place where this orchid occurs, after a lapse of many years!

Perhaps more remarkable than anything else are such cases as the Yucca and the Yucca-moth or the Fig-wasp and the Fig.

The Yucca is a fine lily-like plant resembling the Aloes in general appearance. A particular sort of moth lives entirely upon the Yucca. When the flowers open, the mother-moth kneads up a ball of pollen and places an egg inside. This ball she thrusts down the style into the ovary of the flower. There a grub develops from the egg and eats the pollen, yet some of this pollen fertilizes the young seeds. If Yuccas died out the moth would be exterminated. If the moths were destroyed, no Yuccas would ever set their seed!

The Fig has two sorts of flower. The one (caprifig) produces only male or pollen-yielding flowers. The other is 80 the true edible fig. Inside the caprifig are the grubs of the fig-wasp, which rejoice in the name of *Blastophaga grossorum*. When grown up these force their way out of the caprifig and, flying to the true fig, the mother-wasp lays her eggs in certain flowers which have been apparently specially modified for the purpose. At the same time she covers the ordinary flowers with pollen from the caprifig. Her progeny return to the caprifig. Here again the future of a valuable fruit-tree is absolutely bound up with the fortunes of a tiny and in no way attractive wasp!

Another very remarkable case is that of those flowers (Stapelia, etc.), which in colour and general marking closely resemble decaying meat or other objectionable substances. Very often the smell of such flowers is exceedingly strong, and resembles the ordinary smell of putrid matter. In one case an artist employed to paint the flower had to use a glass bell, which was put over it. He could only lift it for a second or two at intervals in order to see the exact colour, before the horrible odour obliged him to cover it over again. Blowflies and others, which are in the habit of resorting to such substances, seek out these flowers in great numbers and lay their eggs upon them. In so doing they carry the pollen.

There are certain fungi which have quite as horrible a smell, and some of them also resemble decaying animal matter. These are most eagerly sought out by the same blow-and other flies (bright green lucilias, yellow-brown scatophagas, bluebottles, etc.). But in the case of these fungi it is the spores, not pollen, which is carried by the insect.

The effect of this flowery sort of life is abundantly evident in the structure of the insects themselves. Their mouth has been most wonderfully modified into a complex sucking apparatus; their legs have been transformed to act as pollen-carrying baskets, and the habits and tastes of the insects have been modified in the most extraordinary way.

Perhaps also the association of bright colours with a very pleasant sensation—that of a full, satisfying meal—has raised the artistic sensibilities of butterflies, sunbirds, humming birds, etc. For certainly these flower-haunting birds and butterflies are remarkable for their brilliant colouring. This has probably been brought about by the preference of the females for the most brilliantly coloured male butterflies and humming birds.

At any rate bright reds and blues are common to both bird or insect and to the flowers that they frequent. But the most curious point of this whole question lies in the fact that human beings of all grades, South Sea Islanders, the Ancient Greeks, Peruvians, Japanese, Romans, as well as the

Parisians and Londoners of to-day, appreciate the beauty of colouring and grace of form which are so obvious in the world of flowers.

Yet man has had nothing whatever to do with the selection of either these colours or shapes. Many of those which he considers most precious (such as the weird, spotted, and outlandish Orchids of Madagascar and South America) have very likely scarcely ever been seen by man at all. It is to the artistic eye of the honey-bee, bumble-bee, butterfly, and of the humming bird and sunbird, that we owe these exquisite colours. The grace and beauty of outline probably depend upon their perfect symmetry and on the perfect suitability of every curve to its purpose.

Therefore it seems that the eyes of man, whether savage or civilized, are pleased and comforted by these same colours that delight the little brains of insects and birds.

This is indeed a mysterious fact.

CHAPTER VI. ON UNDERGROUND LIFE.

MOTHER-EARTH—QUARRIES AND CHALK-PITS—WANDERING ATOMS—THE SOIL OR DIRT—POPULATIONS OF WORMS, BIRDS, GERMS—FAIRY RINGS—ROOTS MILES LONG—HOW ROOTS FIND THEIR WAY—HOW THEY DO THE RIGHT THING AND SEEK ONLY WHAT IS GOOD FOR THEM—ROOT VERSUS STONES—ROOTS WHICH HAUL BULBS ABOUT—BISHOPSWEED—WILD GARLIC—DANDELION, PLANTAIN—SOLOMON'S SEAL—ROOTS THROWING DOWN WALLS—STRENGTH OF A SEEDLING ROOT.

THE word "Adam" means red earth. Poets and essayists still regularly write about Mother-Earth and, in so doing, admit one of the most interesting and wonderful facts in Nature.

If you go to some quarry or cliff where a section has been cut, laying bare the original rock below; then (with Hugh Miller) you may reflect on the extraordinary value of those few inches of soil which support the growth of all our trees and of all our cultivated plants.

It is probable that plant-roots *never* go deeper than about thirty feet. All our food, our energy, and activity depend therefore on this thinnest surface-layer of an earth which is 8000 miles in diameter. But in most places the depth of true soil is far less than thirty feet, generally it is not more than thirty inches, and by far the most valuable part of it is a very thin layer five or six inches thick.

It is in this true soil that the roots gain their nourishment, and not only roots, for whole populations of worms, of germs, of insects, even of birds and the higher animals, live upon it. To it return the dead leaves, the bodies of dead insects, and waste products of all kinds. Within it, they are broken to pieces and worked up again by the roots of other plants in order to form new leaves, new insects, and food for bird and beast. Just as in engine-works, you may see old engines, wheels, and scrap-iron being smashed into pieces; they are melted down and again worked up into engines of some improved design.

Mares are driven at the gallop round the circle, and so beat the corn out of the ear with their hoofs. They do this for twenty minutes at a time, and are then made to go the other way round.

On a chalk-cliff, which dates from the long-distant Cretaceous period, the entire thickness formed by the yearly work of plants for millions and millions of years is often less than a foot in depth, and probably only four to five inches are true soil.

But this is an exceptionally thin stratum, although it is capable of producing rich turf, fat snails, and excellent mutton. In peat-mosses and in those buried forests which form the coalfields, vegetable matter may accumulate in deposits of thirty feet of coal. Yet these stores of carbonaceous matter seem to be at first sight miserly and selfish, at least from a vegetable point of view.

They resemble the gold and silver withdrawn from circulation in the world by some Hindoo miser and buried deep within the earth. Yet somebody is pretty certain to find out and make use of such stores eventually.

In the case of the peat and coalfields, an animal of sufficient intelligence to utilize them has already been produced, and now they are used by man as fuel.

It is very important to remember that the soil is a sort of last home to which the particles of carbon, of nitrate, and minerals always return after their wanderings in the bodies of plants, of insects, or of other animals. They probably rest but a short time before they again set off on new adventures.

One might say the same of the water, and of the carbonic acid gas and oxygen of the atmosphere, for the water, falling as rain upon the earth, trickles down to the underground water-level. Then it immediately begins to rise up between the particles of earth and is promptly caught and sucked in by the roots, only to be again given out by their leaves. The carbonic acid gas and oxygen also are always entering and leaving the foliage. Even the nitrogen of the air is not left alone in the atmosphere. There are small germs in the soil which are able to get hold of it and make it into valuable nitrates.

More curious still is the fact that electric charges can be used to change the comparatively useless air-nitrogen into useful manures. Probably the farmer will some day make his own nitrates by electricity.

The structure of the soil or earth is a most interesting and romantic part of botany. It is true that a "radical" disposition is necessary if one is to go to the root of the matter, but, unless we do this, it is impossible to realize the romance of roots.

Down below is the unaltered rock, sand, or clay. Next above it comes the subsoil, which consists of fragments of the rock below, or of sand, clay, etc., more or less altered by deep-going roots. Even in this subsoil, bacteria or germs may be at work, and the burrows of worms and insects often extend to it. Next above the subsoil comes the true soil; there is plenty of the stones, soil, sand, or whatever it may be that constitutes the subsoil, but its richness consists in its contents of valuable minerals, and especially of broken-up leaves, corpses of insects, and manure. Above this true soil are first the leaf-mould of two years ago, then that of the year before last, and *on the top* is the leaf-mould and other decayed products of last winter.

All these upper layers are full of life and activity, which probably goes on vigorously all the year round.

The population of worms is especially important. The worm is a voracious and gluttonous creature: it is for ever swallowing bits of leaves and rich soil. Inside its body there are lime-glands which act upon the vegetable food and improve its quality as manure. The worm comes up to the surface at night or early morning and leaves the worm-casts upon it. The rain then washes the rich, finely-divided matter of the casts down into the soil again. It is said that there are about 160,000 worms at work in an acre of good soil. Yet their life is full of danger. A keen-eyed population of blackbirds, thrushes, starlings, peewits (plover), and partridges are always watching for and preying upon the poor worm. Even in his burrows, which may be six feet deep, he is not safe, for the mole (*moudiewarp*) is also both very hungry and very active, and delights in eating him.

In the soil also and even deeper in the subsoil are many insects; some hibernate in the winter, and at other times actively gnaw the roots of plants or devour dead leaves and twigs (see Chapter xxiii.). Thus there are many burrows and holes, so that there is no want of air in the soil, which is indeed necessary both for these creatures and also for the roots of the plants.

Rain comes down through the soil, carrying with it carbonic acid, mineral salts, and also germs or bacteria, which form perhaps the most important population of all.

No work could be carried on without their help; it is bacteria which, at every stage of decay, assist in breaking up leaves, twigs, insects' bodies, worm-casts, and other manures. The way in which they work is too difficult to explain here, but to get an idea of the romance of the underground world one must try to picture to oneself these swarms and myriads of germs and bacteria all incessantly and busily engaged at their several duties. In the uppermost layers there are probably in a single cubic inch of good soil from 54,000,000 to 400,000,000 of these microbes. Many are absolutely necessary to the harvest; a few may be of little importance, but there are sure to be some of those dangerous sorts which might devastate a continent with disease in a single summer.

There are also quantities of other fungi. The fairy rings which one sees year after year in widening circles of bright, fresh green are the work, not of fairy footsteps, but of an underground fungus (*Marasmius oreades* and others). Its threads are thin, white, and delicate; they attack the roots of grasses, etc., on the outer side of the ring. It is therefore on this outer side yellow, dry, and more or less withered. On the inner side, however, the grass is luxuriant and of a rich bright green. Here the fungus has died off, and its remains, as well as those of the plants which it destroyed, form a rich manure for the new grass following on its track. Every year the ring widens; at a certain time in summer one sees the irregular line of mushroom-like fungi which are formed by the destructive underground absorbing threads. This, however, is but one of the underground fungi. There are many kinds; some are useful, others are very destructive.

Upon the upper surface of the soil there falls not only rain, but another sort of rain consisting of seeds, dead leaves, insects' bodies, fungus spores, bacteria, and dust.

Every year when the ploughman turns the sod there is a revolution in the whole of these populations.

So far nothing has been said about the roots themselves, which penetrate, explore, and exploit all these layers of dead leaves, soil, and subsoil.

The length of roots produced is very much greater than any one would suppose. A one-year-old Scotch fir seedling when grown in sand produced in a season a total length (branches, etc.) of no less than thirty-six feet of root. The total surface of this root system was estimated to be about twenty-three square inches. This little Scotch fir after six months' growth was laying under contribution a cone of earth twenty to thirty inches deep and with a surface of 222 square inches. In certain kinds of corn the same author estimated the total length of the roots as from 1500 to 1800 feet. S. Clark estimated the length of the roots of a large cucumber plant as amounting to 25,000 yards (fifteen miles), and made out that it was occupying a whole cubic yard of ground.

Clover roots are said to go down to depths of six or nine feet, but many weeds go deeper still. Coltsfoot, for instance, may be found, according to a friend of mine, living at a depth of twenty spades. In Egypt and other places the roots of acacias go down to twenty feet or even further, so that they can tap the water supplies, which are at a great depth.

But a still more extraordinary fact is the manner in which the root-branches arrange to grow in such a way that they search every part of the soil.

The main root in many plants grows straight down, or as nearly as it can do so. Its branches are inclined downwards at a quite definite angle which is often 30°-45° to the surface. Moreover, these branches come off in quite a regular way. Each keeps growing in its own special direction to the east, south-east, or west, or whatever it may be, of its parent root.

Have they some extraordinary sense of the direction of the points of the compass? It is said that if a side root, which is growing, say for instance downwards and westwards, is turned in some other direction, it will after a time resume its original westerly voyage. This fact is a most extraordinary one, if true, but it can scarcely be said that it has been proved, and, as will be shown later, there are other curious facts in the behaviour of roots which might explain the experiment without assuming that roots know the points of the compass.

If one cuts a branch of willow and plants it upside down in the earth, it will very likely take root and grow. Its appearance will be most extraordinary, for the roots will grow downwards, whilst the branches, instead of growing in the direction of the old branches, turn round and grow upwards.

Why do roots generally grow downwards? The fact is so familiar that the difficulty of answering does not, at first sight, seem so great as it really is.

Pfeffer, the great physiologist, has the following interesting comparison. Suppose a man is trying to find his way in the dark, then a single lingering ray of light gives him an impulse to walk towards it. So our root, also in the dark, feels the pull of gravity and endeavours to grow downwards. Others have compared the direction of gravity to the sailor's compass, and suppose that the root is guided in the same sort of way.

But a young, vigorous root making or forcing its way in darkness through stones and heavy earth is a most interesting and fascinating study.

There are the most extraordinary coincidences in its behaviour. It has the property of always doing exactly the right thing in any emergency.

It is of course intended to keep below the ground and in the dark. So we find that if roots are uncovered, they will turn away from the light and burrow into the earth again. They avoid light just as a worm would do.

Roots are of course intended to absorb or suck in water. If there is a drain in the soil or a place where water collects, the roots will grow towards that place. Very often they form a dense spongy mass of fibres which may almost choke the drain. Along a riverside one can often find great fibrous masses of tree roots near the water. But how does the root learn that the water is there and turn away from its original track to find it? It certainly does so!

Then again, Herr Lilienfeld has recently shown that roots seem able to turn away from poisonous materials in the soil and to seek out and grow towards valuable and nutritious substances. He found that peas, beans, sunflower, and other roots were very sensitive to different substances in the soil, and were directly attracted by what was good for them and turned aside from what was unwholesome.

This property and the power of growing towards water probably explain the mysterious sense of direction alluded to above, for roots will take a line which has not been exhausted by their neighbours.

But of all these wonderful properties, the most remarkable is the way in which roots find their way past stones and other obstacles in the soil. They insinuate themselves into winding cracks and crawl round stones with an ingenuity that makes one wonder if they can possibly be without some sort of intelligence.

It is the very tip or end of the young root that seems to be responsible; for if, in the course of its journeyings underground, it should strike a stone or something hard, the root does not grow on and flatten itself.

But some sort of message is sent back from the tip to the growing part which is a short distance behind it. After this message has been received, the growing part begins to curve sideways, so that the tip is brought clear of the obstacle and can probably proceed triumphantly upon its way. The inexplicable part is that the growing part which curves has never been touched at all, but simply answers to the message from the tip.

This is perhaps the most reasonable and intelligent behaviour found in the whole vegetable world, and it is not surprising that Darwin compared the root-tip to a brain.

These extraordinary responses fill one with astonishment, but there are others still more interesting and remarkable. It will be remembered that we have already shown how different the soil is at different levels. The subsoil, soil, and uppermost layers are all quite different from one another.

This may explain why it is that many plants seem to prefer to develop their roots at one particular depth below the surface. Not only so, but they find their own favourite level in the most persevering way.

If, for instance, you sow a barley-corn at too great a depth, the seed germinates and forms a few roots, but it immediately sends out a stem which grows upwards towards the light. As soon as this stem has reached the proper place, which is just below the surface, there is an enormous development of roots, which begin to search and explore their favourite stratum of soil.

In some few cases one can see in a dim sort of way the reason for the level which certain plants prefer. Thus the underground stems of the common Thistle, which are very long and fleshy, are found just a few inches below the level usually reached by plough or spade. This makes it very difficult to tear them out. Even if grubbers with long spikes which reach as deep as these buried stems are driven through the ground, it generally happens that the stems are only cut in pieces and not dragged up. These hardy weeds are not much injured by little accidents of this kind, for each separate bit will form upright thistle stems next year. In fact if one cuts this fleshy subterranean runner of the Thistle into pieces a quarter of an inch long, each piece will probably become a Thistle.

Sometimes indeed these weeds are carried from one field to another by pieces of them sticking in the very machines which are used to eradicate them.

The Bishopsweed is one of the hardest cases. The writer was once ambitious enough to try to dig up an entire plant of this horrid weed. The first foot or so revealed no sign of the end of the branching runners, and it was not until a hole about four feet deep and five feet across had been excavated that there was any sign of an end to the plant.

When it was at last removed, the original deeply buried stem was found to give off branches which again branched in a most complicated manner, until almost every green shoot of Bishopsweed within a space six feet in diameter was seen to be really a branch of this one original plant! So to eradicate the plant it would have been necessary to dig over the whole garden to a depth of at least five or six feet.

How did the stem get down to such a depth below the surface? This is one of the most curious stories in plant life, and the process which we shall now try to describe has only been explained within the last few years.

The seed of the Wild Garlic (*Allium ursinum*) lies at first upon the surface of the ground, but it is soon buried by a growth of the stalk of the seed-leaf, which pushes the germ down below the earth. As soon as it is buried, roots are formed and pass obliquely downwards, where they become fixed by forming root-hairs all round themselves. These root-hairs round every root hold its tip firmly in the earth; then these same roots contract or shorten, which of course hauls down the root a little deeper in the earth. One might compare it to a few men hauling down a balloon by ropes attached to the car. About September to November, roots of quite a different character are formed; these explore the surrounding soil and gather in food and moisture.

Then the roots rest during the winter, when the buds and young leaves are being formed. In April the buds begin to push out their leaves and a new ring of roots appear. These April roots are quite different from the September ones. They again fix themselves firmly and then contract, becoming fully a third shorter than they were originally. The bulb is dragged down still deeper below the surface. It flowers in May and fruits in June and July. Then in September the same series of operations begins again. The process goes on until the plant is three to five inches below the ground.

It follows from all this, that every year the roots find new ground to explore and utilize. Nor is the Wild Garlic at all exceptional in this respect. A great many plants have roots which contract and drag the bulb or stem after them deeper into the earth. Something of the same sort happens, for instance, to Bramble branches. They arch or droop over, when growing, so that the end touches the earth. On the underside of the tip, as soon as it begins to rest on the ground, roots are formed. These roots make their way into the ground, and then, when fixed, they shorten or contract, so that the end of the branch is dragged down to a depth of several inches. After this has happened the old branch generally dies away, and a young, vigorous Bramble develops from its buried tip.

Raspberry branches also are often buried; their roots become coiled or rolled in a very curious manner. The end of the root becomes firmly attached in the soil, and then the rest of it revolves like a tendril so as to draw the stem deeper into the earth.

On any ordinary roadside in the country one is sure to find the rosettes of the common Dandelion and of the Rats-tail Plantain (*Plantago major*). These are two of the most interesting plants in the world, although they are vulgarly common. How is it that their leaves are always at the level of the ground? The stem is always growing upwards; every year fresh circles of leaves are formed above the older ones. Yet the crown of the stem is never so much raised up above the ground that the toe of a boot would be likely to knock it off. It is always kept so deep in the earth, that it is by no means easy to kick or "howk" the crown out of the ground.

The Dandelion root contracts very strongly at the end of the season, and by this shortening or contraction keeps its leaves just at the soil level. The Plantain sends out about forty to sixty oblique downward-growing roots, which fix themselves in the soil by throwing out branch roots. These forty to sixty roots are at first about ten inches long, but, as soon as they are firmly attached, they contract, and pull the stem with its crown of leaves about one-third of an inch deeper. This is just enough to keep the leaves flat on the ground and to prevent any possible injury from passers-by.

So that in finding their favourite level in the soil, plants are often pulled or hauled about by the roots. But they are not always moved by the roots. Even though buried in darkness, they seem able in some way to tell when they are in the most favourable position.

Every gardener knows that Autumn Crocus and other bulbs do not remain in the same position. They wander below ground in a curious and inexplicable fashion.

The Solomon's Seal has an underground, fleshy stem, which prefers to grow at a definite depth. If it is planted close to the surface, then the point of the next year's little fleshy bud turns downwards; next year it again turns downwards, and so on every year, until the stem has reached its proper depth. Then it grows horizontally. Similarly, if it is planted too deep it grows upwards.

Thus if one wishes to realize the underground life of plants, one must picture to oneself:—

1. The usual descending roots, whose system of branching may be compared to the ordinary branching above ground. It is often not unlike the reflection in water of the tree itself, such as one might see on a fine winter's day along the shore of some still lake.

2. The bold, exploring, horizontal runners of Couchgrass, Thistle, Bishopsweed, etc., vigorously pushing their way at a depth too great for the gardener's spade.

3. All sorts of bulbs, runners, and roots being slowly hauled or dragged about till they get into exactly the right position, but never remaining for two years in exactly the same place. All have their favourite depth—

Herb Paris	2/3 to 1-3/4	inches deep.
Solomon's Seal	1-1/3 to 2-1/3	" "
Cuckoo Pint (*Arum maculatum*)	2 to 4	" "
Colchicum (Autumn Crocus)	3-1/3 to 5-1/3	" "
Asparagus	6-3/8 to 13-1/8	" "

The water evaporating on the surface of the soil must, as it rises from the permanent water-level below, pass the gauntlet of all these thirsty rootlets and their hairs. Tree-roots will be ready to intercept it at ten feet depth, many herbaceous plants will suck it in at depths of five to six feet, and in the upper layers of soil it will have to pass root-system after root-system from Asparagus to Paris, so that very little will be lost.

Perhaps of more importance are the bacteria-germs, and dissolved mineral salts in the rainwater as it trickles down from the surface. The soil particle acts as a filter: at every inch of the descent some of the bacteria and salts will be left, so that by the time the level of Asparagus has been reached there will be exceedingly few, and the water is comparatively speaking pure. The effect of this vigorous underground life is often visible on the surface. Roots, and particularly tree-roots, are often extraordinarily strong. Kerner, in his invaluable *Natural History of Plants,* has a beautiful picture of a young larch tree which had grown in a fissure of a huge boulder.

In attempting to grow, the root had forced up part of this stone. It was estimated that it had lifted a weight of 3000 lb., though it was only some ten inches in diameter.

Along a dry-stone wall, or even near houses, the growth of tree-roots very often damages the entire wall, which may be entirely overthrown if the tree is too near. The force of the growth of the roots is so great that even a six-foot stone wall cannot keep them down.

Quite a young seedling root, in forcing itself through the soil, may exercise a pressure of two-thirds to four-fifths of a pound!

This is of course necessary, if one remembers that it has to drive itself through the earth, pushing aside and compressing the earth particles along its course.

CHAPTER VII. HIGH MOUNTAINS, ARCTIC SNOWS.

THE LIFE OF A CHERRY TREE—CHERRIES IN MARCH—FLOWERING OF GORSE—CHICKWEED'S DESCENDANTS—FOREST FIRES IN AFRICA—SPRING PASSING FROM ITALY TO THE FROZEN NORTH—LIFE IN THE ARCTIC—DWARFS—SNOW-MELTING SOLDANELLAS—HIGHLAND ARCTIC-ALPINE PLANTS—THEIR HISTORY—ARCTIC BRITAIN—EDELWEISS—AN ALPINE GARDEN.

IT is impossible to understand and very difficult to explain the sort of life and consciousness which is enjoyed by plants.

That they do live is obvious; we know instinctively that they enjoy fine weather in summer and gentle showers in spring, but we cannot prove it.

Much of a plant's life is concealed and hidden from us. Even the few explanations which have been given by certain observers are by no means generally accepted.

This is true even as regards the case of the Cherry tree, which has been experimented with, and fought over and argued about by botanists, and yet we only know a very little about its inner life.

When the leaves fall in autumn, next season's buds are already formed and are then about one-eighth of their full size. At this time the tree contains enormous quantities of food-stores, for the whole season's work of the leaves has been accumulating until this moment. During the long winter's "sleep" the tree is by no means at rest. It is arranging and packing up those stores in the safest place and in the most convenient form.

Just as a bear, before it retires to sleep during the winter, takes care to get as fat as possible, so the Cherry turns its starch to fat, and stores it away in the innermost and least exposed parts of the tree, that is in the central wood. As soon as the winter ends, and indeed *before* it has ended, preparations are beginning for the great moment of the year. For weeks there is a slow, gradual, almost imperceptible growth of the buds, then they develop with a rush, and in six to ten days double or treble their weight. Then comes the supreme moment, for the flower-buds suddenly burst open and the Cherry is in active and vigorous bloom and covered all over with exquisite blossoms. All last year's fats and starches are rapidly used up. Very soon the young leaves are beginning to make sugar and other food, which give some help during the ripening of the fruit.

The flowers are actively at work. One of our usual misconceptions as to the nature of a flower is that it is an emblem of peace, of restful enjoyment, of serene contemplation of its own beauty. That is very far from being the truth. The petals are actively, vigorously working. If one could take the pulse of a petal, which shows the rapidity of its breathing, one would find that it is twice as fast as that of the leaf. The work of changing water into vapour and pouring it out goes on three times as quickly in the petals (as compared with the leaves). Moreover their temperature is higher, and often distinctly above that of the atmosphere.

This feverish activity of the flowers themselves is matched by the hurrying crowds of excited and exhilarated insects which are searching every blossom.

No wonder that the Japanese Prime Minister, in the midst of their great and famous war, invited the whole cabinet to spend an afternoon watching the cherry trees in bloom!

From the blossom of the springtime all through summer and autumn follows one continuous spell of hard work. Day after day an endless stream of food is entering the stem; night after night it is condensed and arranged and repacked, until, when the leaves fall, the period of slow and quiet preparation begins again.

Under certain conditions it is possible for gardeners to modify the life of a cherry, and to make it bloom much earlier, but this is only possible within well-defined limits. It is no use trying to force it to bloom before January. It *must* have a quiet time after summer. But by beginning in January and by very carefully managing the temperature, it can be made to produce fruit quite early in the year.

The following account is given to show how very carefully gardeners have to work when they upset the ordinary course of Nature's events. The plant is taken into a greenhouse, and the temperature kept as follows:—

	Day Temperature.	Night Temperature.
First week	48°-50° F.	41°-45° F.
Second week	50°-53° F.	45°-48° F.
Third week	53°-59° F.	48°-51° F.
Till flowering	59°-64° F.	51°-57° F.
Flowering period	46°-53° F.(!)	43°-50° F.(!)
After flowering	59°-64° F.	51°-57° F.
During development of stone	53°-59° F.(!)	48°-51° F.(!)
After development of stone	61°-66° F.	53°-59° F.
Ripening of fruit	68°-70° F.	59°-63° F.

Not merely strong, forcing heat, but a little judicious cold, is necessary to get out the flowers and to ripen the fruit.

Most flowers have very much the same general history as the cherry, but it must not be supposed that they are all alike. The differences are very interesting and curious.

Thus, for example, plants of our common Gorse, furze, or whin may be found in bloom at almost every season of the year. There are at least four seasons when there is that tremendous display of golden blossom which made the great Linnæus fall on his knees and burst into tears. These are about the 22nd March, 24th May, 15th August, and 21st November; yet there are enough odd flowers blooming in almost every month to give some cause for the saying, "The gorse is out of bloom when kissing is out of favour." The last practice, though uncleanly and dangerous, not only on general grounds, but on account of bacterial germs which may be transferred, has been authoritatively condemned in the United States, but it is still more or less popular in other countries at all seasons.

The Chickweed and some other of our annual weeds show a hardy disregard of climate. Its seeds germinate and grow at any time, so that flowers and seeds can be formed whenever there is a spell of favourable weather. Now one chickweed can produce 3000 seeds. Suppose that there are only five generations in the year, which is a very low estimate. Then one seed of chickweed might produce 3000 × 3000 × 3000 × 3000 individuals in one season!

Other plants show much the same tendency. In fine warm autumns a great many annuals bloom a second time. It is on record that forty-four spring species bloomed in one warm November. At the Cape and in other warm climates many of our annuals do not die at the end of autumn, but go on growing. They become perennial.

It is even possible to make a Tree Mignonette by pinching off the flower-buds, though this plant is usually an annual.

In fact plants are not absolutely confined to one rigid scheme, but they can alter and modify their blooming time if they find it convenient to do so. In the Mediterranean some blossom in early spring and others in late autumn, whilst in the dry, hot, and dusty summer very few flower.

In Central Africa during the dry season forest fires are by no means rare. The trees are scattered, and the ground is only covered by dried and withered grasses and sedges. One sees in the distance a rolling cloud of smoke, and soon one comes to a line of flame. It is not dangerous, not even very impressive, for a jump of three feet carries you over the flame and on to a desolate wilderness of black cinders, out of which stand up the scorched trunks and half-burnt branches of gaunt, naked trees. A day or two afterwards, bright blue and white and yellow flowers break out of those scorched branches and also from the ground.

It is difficult to understand why this happens, but certainly it is good for the flowers, which can be seen by insects from a long distance.

But these are unusual cases. Generally the warm breath of spring wakes up the bulbs and buds, and one after another has its moment of flowering.

Spring travels towards the North Pole at an average rate of four miles a day.

A pedestrian visiting Italy in the end of January might follow the spring northwards, and if he wished to accompany it all the way, it would be quite possible to do so without exceeding an ordinary day's march. He would have to reach North Germany by the end of March, Sweden in May, and by the end of June and July would find spring beginning in the desolate Arctic regions.

Of course the presence of mountains would make this tour a little difficult and devious, but still it is quite a possible undertaking. It would be very interesting, for he would be able to watch the cold and frost and chilliness of winter disappearing as the sun's rays thaw out a greater and greater extent of the cold and frozen North.

The life of an Arctic plant is truly set in the midst of many and great dangers.

For 250 days the ground is hard frozen and the temperature *never* above the freezing-point. About the end of May it begins to rise a little, but the plant has to crowd the whole of its life, its flowers, fruits, and seeds, into the space of two months!

About the 23rd to the 29th June the first flower appears, then follows strong, active growth in uninterrupted sunshine during July and August. The flowers are brilliant in colour and richly produced. The tiny dwarf Arctic plants are covered all over with blue or golden yellow or white blossoms. All is in full activity and luxuriance. Then suddenly, in a night, the icy grasp of winter falls upon them.

Hard-frozen flowers, buds, and ripening fruits remain chilled and incapable of life from the 30th August until the end of May.

Of course, under such conditions, these hardy and vigorous little plants cannot become trees or shrubs. To show the effect of the climate upon them, a few British plants which are also Arctic may be compared.

	In Great Britain.	In the Arctic Regions.
Matweed (*Matricaria inodora*)	6 in. to 1 ft. high	2 in.
Goldenrod	1-2 ft.	3-4 in.

(*Solidago virgaurea*)		
Red Rattle (*Pedicularis palustris*)	6 in. to 1 ft.	2-3 in.
Mugwort (*Artemisia vulgaris*)	2 to 4 ft.	4-5 in.
Willow-herb (*Epilobium palustris*)	1 to 2 ft.	2 in.
Grass of Parnassus (*Parnassia palustris*)	6 in. to 1 ft.	1 in.

These wretched little dwarfs seem, however, to have pretty long lives, and, as we have said, deck themselves in the most gaudy colours every summer.

In the Alps of Switzerland and other temperate countries, the flowering season is also a very short one and soon over. It is often not more than six weeks, yet in that short time the rich blue of the Gentian, the Alpine Roses, Soldanellas, Campanulas, and many others make some of these grass slopes high up in the mountains a perfect garden of loveliness.

Sometimes in passing over the snowfields of Switzerland just before spring, one notices the pretty violet flowers of the Soldanella swaying to and fro in the wind above the unmelted snow. One does occasionally see in this country the Snowdrop in the midst of snow, but then it has fallen after the Snowdrop had blossomed.

The Alpine Soldanella flowers whilst the earth is still covered. It begins as soon as the ground below the snow is thawed. Each little developing flower-stalk melts out a grotto in the snow above itself, and so bores, thawing its way up into the air above. It has already been mentioned that, inside a flower, the temperature is often higher than the surrounding air. It is this higher temperature of the flower which thaws a little dome or grotto in the snow above the head of the flower. When a flock of sheep are covered by a snowdrift, a similar hollow is formed above them by their breath and the high temperature of their bodies: they often seem indeed to be little or none the worse for being buried. The Soldanella melts its way in just the same manner.

In this country we have no such magnificent chain of mountains as the Alps, and yet we find on the Scotch and Welsh mountains quite a number of real alpines.

There are, for instance, such flowers as Sea-pink (*Armeria*), Sea Plantain (*Plantago maritima*), Scurvy-grass, and others, which can be found on windy, desolate gullies and corries high up on the Highland hills, and which also occur on the sea-coast, but *never between the seashore and the tops of the mountains*. You might search every field, every moor, and every riverside throughout the country, but you would not discover those three plants anywhere between the seashore and the summits.

At first sight it seems quite impossible to explain why this should be the case. But all those three plants are found in the Arctic regions, and the explanation is in reality quite simple.

At one time the shores of England and Scotland formed part of the Arctic regions. Ice and snow covered the hills and mountains; huge glaciers occupied the valleys and flowed over the lowlands, plastering the low grounds with clay which they dragged underneath them, and polishing and scratching any exposed rocks.

When the ice began to melt away and left free "berg battered beaches" and "boulder-hatched hills," Lincolnshire and Yorkshire must have been like the Antarctic regions in those days. This is how Dr. Louis Bernacchi describes the Antarctic continent:—

"The scene before us looked inexpressibly desolate.... No token of vitality anywhere, nothing to be seen on the steep slopes of the mountains but rock and ice.... Gravel and pebbles were heaped up in mounds and ridges. In some places these ridges coalesced so as to form basin-shaped hollows. Bleached remains of thousands of penguins were scattered all over the platform, mostly young birds that had succumbed to the severity of the climate."

Great Britain must have been just as savage and desolate when these hardy little Arctic plants colonized the shingles and rooted themselves amongst the rocks.

They covered not only the seashore, but they probably made a settlement wherever rock or land of any kind was exposed. These original settlers have had three bands of descendants. One band has remained ever since on the seashore of Great Britain; another set gradually travelled northwards. As the ice melted away, leaving the land bare, first in Denmark, then in Norway, and finally in Greenland, this second set followed it, until now we find them far to the northward, populating the Arctic regions of to-day just as they did those of Britain in the Great Ice Age.

The third set of descendants would at first cover all the land and rocks of the lower hills and valleys near the sea; then as the ice and snow melted and exposed the higher mountain sides, they would climb the hills and eventually reach the exposed summits where they are now living. There they find themselves in an impossible, savage sort of climate, in which they alone are able to exist. Violent storms, drenching mist, scorching sunshine (when the rocks become so hot that it is almost impossible to touch them), rainstorms and months of snow and hard frost, cannot kill Scurvy-grass, Seathrift, or Plantain, but there are few other plants which can stand such conditions. Lower down on the flanks of the hills and in the valleys, they have long since been dispossessed of the rich and fertile lands by plants which can grow more rapidly and luxuriantly.

The little Alpine Creeping and Least Willows, for instance, some of which get up to 3980 feet in Breadalbane, are mere dwarfs only a few inches high, and totally different from their allies in the fertile lowlands, which are trees eighty to ninety feet high.

Some of the Alpine plants which also occur in the Arctic regions have not even been able to survive by the seaside in Great Britain. Their nearest allies are in the Norwegian mountains.

It would be impossible even for shrubs to stand the violent winds and snowstorms of these summits. Alpine plants are generally low-growing mats. They are also often clothed all over in cottonwool, such as the Edelweiss. This probably keeps them from losing too much water during the dry season, when the rocks on which they grow are strongly heated by the sunlight.

Yet, like the Arctic plants, they have rich, deep, and brilliant colours.

A queer point is that they have got so accustomed to this stormy and perilous existence that it is extremely difficult to grow them in a garden. Like mountaineers, they dwindle and pine away in the richer soil and softer air of the low grounds.

To make an Alpine garden, rocks and stones must be arranged with pockets and hollows, like natural crevices and basins, between them. Rich leaf-mould must be placed in these hollows. There must be good drainage, and as much sunlight as one can possibly get.

CHAPTER VIII. SCRUB.

FAMOUS COUNTRIES WHICH WERE COVERED BY IT—TREES WHICH ARE COLONIZING THE DESERT—ACACIA SCRUB IN EAST AFRICA, GAME AND LIONS—BATTLE BETWEEN ACACIA AND CAMELS, ETC.—AUSTRALIAN HALF-DESERTS—EXPLORERS' FATE—QUEEN HATASU AND THE FIRST GEOGRAPHICAL EXPEDITION RECORDED—FRANKINCENSE, MYRRH, GUMS, AND ODOROUS RESINS—MANNA—LADANUM—BURNING BUSH—OLIVES, ORANGES, AND PERFUME FARMS—STORY OF ROSES—BULGARIAN ATTAR OF ROSES—HOW POMADE IS MADE—CUTTING DOWN OF FORESTS AND MOHAMMED.

A scrub or Half-desert does not seem at first sight to be in the least interesting.

But if one remembers such places as Cordoba, Seville, Florence, Genoa, Sicily, Athens, Constantinople, the great cities of Ephesus, Corinth, etc., of St. Paul's Epistles, Persia, Arabia, Palestine, and Carthage, surely the countries which have had such splendid histories deserve a chapter to themselves. What achievements in war, in art, in literature, and in romance are connected with these lands bordering the Mediterranean or fringing the great deserts of Sahara and central Asia!

The animals which belong to such country are also interesting. It is the home of the camel, ass, horse, donkey, not to speak of the giraffe, rhinoceros, gazelle, antelope, zebra, lion, and hyena.

The plants are full of interest too, and some of them are of great importance to man. The Olive, Orange, Fig, Roses, and many perfumes and spice-trees, are natives of scrub. In fact, it is the real centre of all gums, frankincenses, and myrrhs.

As man depends upon plants and animals, and as animals also are dependent on the plant world, it is the climate which really is responsible for everything.

The world of plants is entirely and exactly regulated by the character of the climate. What, then, is the climate of scrub?

Those countries enjoy brilliant sunshine, cloudless skies, and yet there is sufficient rain to permit of irrigation and to prevent the unmitigated desolation of the desert. When, as has happened in many of these famous lands, the forests have been cut down and the aqueducts have been neglected, they become arid, dry, and almost useless. But when carefully and industriously worked, as they were in the days of Greece, Carthage, and Rome, they produce results which will for ever live in the history of the world.

The meaning of such half-desert climates and of the scrub which covers them has been already suggested.

The scrub is trying to occupy the desert.

If one takes the sternwheel steamer at the First Cataract of the Nile and passes southwards, the desolation of black rock and "honey-coloured" sand of the Libyan Desert is at first unbroken. But here and there the thorny trees of the "Seyal". Acacia show the beginnings of a scrub region. Much further to the south, those acacias and others become great forests which extend all along the south of the Sahara Desert and furnish the valuable gums of the Soudan.

If one passes southward through this forest of acacias, it alters in character. The trees become taller, closer together, and climbing plants and undergrowth become more frequent. Still further south, one finds the regular tropical forest which is characteristic of the tropics everywhere.

The most interesting part, which is also the richest in big game, is the intermediate zone between the desert and the acacia forest or scrub.

All sorts of transitions are found. Sometimes there are thickets of thorny bushes. Occasionally scattered clumps of woodland alternate with stretches of grass or what looks like grass. Near the desert one finds pioneer acacias dotted singly here and there; these are the scouts or skirmishers of the army of trees which is trying to occupy and colonize the desert.

This explains why this sort of scrub occurs in so many parts of the world. On the European side of the Mediterranean, the dry climate of Spain, the

Riviera, and Greece must no doubt at one time have supported a scrub vegetation. At present it is difficult to tell what this was. There is a sort of scrub called *Maqui* which covers parts especially of Corsica and other Mediterranean countries. In Greece, also, thorny, woody little bushes are very common.

But these are just what the goats, who are fiends from a vegetable point of view, have been unable to destroy. We cannot tell what sort of country revealed itself to the first Phœnicians when they landed in Southern Spain to traffic with the savage inhabitants, or what met the eyes of Ulysses when he made his great voyage to unknown lands.

But there are places in the world where man has never either kept domestic animals or cultivated the soil. Possibly Spain and Sicily in those early days were not unlike parts of British East Africa, such as the Taru Desert between Mombasa and Kibwezi.

The following may give an idea of how this scrub or desert appeared to me.

Gnarled and twisted acacias of all sorts and sizes, usually with bright white bark and a thin, naked appearance, cover the whole country. Amongst these one finds the curious *trees* of Euphorbia. In Britain Euphorbias are little green uninteresting weeds, but here some of them are twenty to thirty feet high, with many slender whip-like branches, but no leaves. Others are exactly like Cactus, and take on weird, candelabra-like shapes. Nobody meddles with them for, if the slightest cut is made in the bark, out pours an acrid, white milk which raises painful blisters, and may even cause blindness if a drop touches the eyes.

Almost all the plants are either covered with thorns or protected by resins, gums, or poisonous secretions.

Between the scrubby trees the soil is dotted over by little tufts of grass or sedge, but these are so far apart that the tint of the landscape is that of the soil.

Game is abundant everywhere. Sometimes it is a small bustard or a persistent, raucous guinea-fowl that affords a chance for a good dinner.

Occasionally a tiny gazelle, the "paa," with large ears, springs out of the thorns and vanishes down the path. I saw footprints of giraffes, and came across ostriches more than once. I also made a persevering attempt to slay a Clarke's gazelle, an animal with enormous ears and a long thin neck.

These long-necked creatures can see far above the usual short thorny bush, and it is exceeding difficult to get near them. Water probably exists under the stony grit soil, but at present one has to be contented with that found in the stagnant pools at Taru, Maungu, etc., which, if not occupied by the decaying remains of a dead antelope, are, as a rule, drinkable.

These acacias are quite well fitted to live in this dry and arid region. Their roots go down to twenty feet or more, so as to reach the deep-seated water supplies.

Their leaves are generally adapted to resist any injury from the strong glare of the sunshine. The gums, already alluded to, are also very important, for any crack or break in the tree is promptly gummed up, and there is no loss of precious water thereby. This gum will also prevent or discourage burrowing and boring insects from getting in; they would, if they tried to do so, become "flies in amber," like those found in fossil resin. The trees are generally provided with strong spines, which guard them from the many grazing animals which try to devour the succulent leaflets.

The fight between the grazing animal and the plant is, in these scrubs and half-deserts, very severe. In Egypt it is said that the whole flora has been entirely altered by the camel and the donkey.

But in this case the battle is unfair. Man keeps those camels, donkeys, and goats. He provides them with water and protects them from lions, leopards, and snakes. In East Africa man has not yet interfered, and the plants probably get the better of the animals. In such places lions, leopards, and hyenas are common. It will be remembered that a lion not very long ago stormed and took charge of a railway station on the line to Uganda, and was only routed with very heavy loss.

There is also some reason to suppose that the antelopes and other creatures do help the plants in their efforts to colonize the Sahara. Their droppings will very greatly improve the soil, and more vigorous thickets and undergrowth will spring up when the soil is improved in this way. Such a vigorous growth of plants will be better able to resist the long eight or nine months' drought, and so help the wood to develop, until perhaps it is too thick, and the trees are too high, for the antelopes to graze upon them. In this manner the Acacia scrub is slowly and painfully colonizing the desert.

It is not only in Africa that one finds these half-deserts or scrub. There is the Brigalow Scrub in Australia, which has a curious silver-grey shimmering appearance on account of the blue-grey sickle-like leaves of the Brigalow Acacia. The foliage casts no shade, for the leaves are flat and thin, and place themselves edgewise to the light, so that there is no danger of the strong light injuring them. Also in Australia is the Mallee Scrub, covering thousands of square miles between the Murray River and the coast. It consists of bushy Eucalyptus, six to twelve feet high. Its monotonous appearance when seen from a small hill is very striking. "Below lies an endless sea of yellow-brown bushes: perhaps far away one may observe the blue outline of some solitary hill or granite peak, but otherwise nothing breaks the monotonous dark-brown horizon. Everything is silent and motionless save perhaps where the scrub-hen utters its complaining cry, or when the wind rustles the stiff eucalyptus twigs."

There is a melancholy interest attaching to both the Mallee and Brigalow, for in them lie the bones of many gallant and persevering explorers. Nor is the East African thorn-tree desert without its victims. The missionary, Dr. Chalmers, was lost near Kibwezi in the Taru Desert.

The ships of the expedition are drawn up along the shores of Punt (in Somaliland), and incense trees are being carried on board. Notice the baboons on board ship, and the rays and sword-fish in the water.

There are a certain number of valuable plants found in these half-deserts or scrubs. Perhaps the earliest geographical expedition of which we have a good account (with illustrations) is that sent by the Egyptian Queen, Hatasu, from Thebes, about three thousand years ago. She built on the Red Sea a fleet of five ships, each able to carry from fifty to seventy people, and sent them to the land of Punt, which was probably Somaliland. The natives lived in round huts built on piles like the ancient lake dwellings. The object of the journey was to obtain incense. No less than thirty-one incense-bushes were dug up with as much earth as possible about their roots, and carried to the ships, where they were placed upright on the deck and covered with an awning to keep off the sun's rays. Whether they did really survive the journey and grow in Egypt is uncertain. Sacks of resin, ebony, cassia, apes, baboons, dogs, leopard-skins, and slaves, as well as gold and silver, were also taken away. The Queen of Punt accompanied them. From her appearance it is not probable that the Queen of Sheba was any relation, although some writers have supposed that Sheba and Punt were the same place.

The whole story is represented in coloured bas-reliefs in the temple at Tel-el-Bahiri, near Thebes.

The incense here alluded to was a very valuable drug in Egypt on account of its use in embalming mummies. Quite a number of gums, resins, and the like, are obtained from Somaliland and similar half-desert countries. The frankincense of the Bible, which may be the incense of Hatasu, is obtained from *Olibanum* produced by various species of Boswellia. In February and March, cuts are made by the incense gatherers in the bark of the trees. Tears of resin soon appear and become dried by the sun over the wound. The best kinds still come from Saba, in Arabia, where the Romans obtained it in the time of Virgil. Besides Olibanum, frankincense contains Galbanum (*Ferula galbaniflua*) and Storax (*Storax officinale*). Equal parts of these were mixed with the horny shield of a certain shell-fish. When the last is burnt, it has a strong pungent odour. The Galbanum is now found in Persia, and Storax in Asia Minor, both half-desert countries. The true Myrrh (*Commiphora myrrha*) is also found in East Africa and South-west Arabia.

The name is supposed to be derived from Myrrha, the daughter of Cinyras, King of Cyprus, who in consequence of a great crime was banished to Arabia and became the tree which bears her name. The myrrh of the Sacred Oracles was used as incense at least 3700 years ago, and it is mentioned by Moses (Genesis xxxvii. 25).

The sovereign of England used always to present gold, frankincense, and myrrh in the Chapel Royal, London, on the feast of the Epiphany, and, strange though it may appear, the symbolic offering is still made each year by our present king.

Balm of Gilead (*Balsamodendron Gileadense*) belongs to scrub or half-desert regions. Cleopatra obtained plants from Jericho for her garden at Heliopolis. The Jews used to sell it regularly to the merchants of Tyre.

It is still valuable, for the essence is worth from £2 to £3 per lb.

The opoponax described by Dioscorides belongs to the Orient. It yields a valuable gum resin, which is much used in perfumery (*Pastinaca opoponax*). It also is obtained by incisions in the bark of the tree.

In fact a very large proportion of these fragrant sweet-smelling substances, Myrrh, Cassia, Bdellium, etc., come from these sunny Eastern lands, which are not exactly deserts but very close to them. Manna, e.g., is obtained from the flowering Ash (*Fraxinus ormus*) in Sicily by transverse incisions being made in the bark, so that the brownish or yellowish viscid juice exudes and hardens on the wound. *Ladanum* is a varnish or gluey coating found on the leaves of *Cistus creticus*, which grows in Crete. In old times the glue was collected from the beards of the goats which had been browsing on the plant. Although this method, no doubt, increased the strength of the perfume, it has been abandoned, and the ladanum is obtained by a "kind of rake with a double row of long leathern straps." The straps take the glue from the leaves. It is used as a perfume in Turkey.

Another very interesting Eastern plant sometimes seen in old-fashioned country gardens in Britain is the "Burning-bush" (*Dictamnus fraxinella*). Like a great many of these half-desert plants, it is full of an acrid, ethereal, odorous substance. On a calm, hot summer's day, this material exudes from the leaves and surrounds the plant with an invisible vaporous atmosphere. Such an atmosphere probably assists in preventing the water from evaporating or being transpired from the leaves.

Now if one places a lighted match a little below the leaves or flowers this vapour catches fire, and there is a display of flames and smoke with little explosions, followed by a strong smell. The plant may be injured if it is set on fire too frequently, but generally does not seem to be any the worse for the experiment.

The Mediterranean is the home of the Myrtle and Olive, of Oranges and Lemons, of Figs and Vines, of Almonds and Raisins, as well as of many other important and interesting plants.

The olive crop in Italy yields about ninety millions of gallons of olive-oil every year. The olives are collected as soon as they become ripe, and are crushed in circular stone troughs with a perpendicular millstone. The paste is then pressed in bags and afterwards clarified by passing through cotton wool. To the eye of a foreigner the white gnarled stems and silver-green foliage of the olive groves are not particularly attractive.

Near Burriana, in Spain, one may walk for miles through the plantations of oranges. The dark-green glossy leaves and golden fruit of the orange make a most beautiful contrast, but the dry, thirsty soil, and the careful way in which the water is regulated and supplied by small gutters, most jealously watched over, make the tourist realize the difficulty of agriculture in so dry and arid a country.

The Myrtle is not a very important plant nowadays, though its berries are still eaten and myrtle wreaths used to be worn by the bride at every wedding. In classical times it was sacred to Venus, but the victors in the Olympian games were also crowned with myrtle, and the magistrates at Athens had the same privilege. It is no longer used as a medicine and for making wine. It is really a native of Persia, but has been introduced to the Levant, Italy, France, and Spain.

It is along the Riviera that one finds a very curious and interesting industry. This is the manufacture of perfumes and essences from the petals of flowers. A great many different flowers are used, such as the Garden Violet, Mignonette (a native of Egypt imported in 1752), Lily of the Valley, Tuberose, "the sweetest flower for scent that grows," Jonquil (*Narcissus jonquilla*), Heliotrope (imported from Peru in 1757), Spanish Jasmine (*J. grandiflorum*), which is a native of Nepaul, and was brought to Europe in 1629, and various Roses.

These Roses have had a long, interesting, and honourable history. No one knows when they were first cultivated. Solomon had his rose-gardens at Jericho. Queen Cleopatra spent some £400 on roses in one day, and Nero is said to have beaten this record by wasting 4,000,000 sesterces (£30,000) in roses for a single banquet.

Rosewater is said to have been first produced by an Arab physician called Rhazés in the tenth century. When Sultan Saladin recovered Jerusalem from the Crusaders in 1187, the pavement and walls of the Mosque of Omar were washed and purified with rosewater. That stout warrior Thibault IV, Count de Brie et de Champagne, brought back roses from Damascus on his return to his native land. That was the origin of the valuable Provence roses. The Lancastrians chose a Provence rose as their badge at the beginning of the Civil Wars of the Roses in England.

Otto of Roses, or the essential oil, was discovered by Princess Nour Jehan at the court of the Great Mogul, and she received as her reward a pearl necklace worth 30,000 rupees. The price of otto of roses seems to have been about £320 per pound in Persia and India when the traveller Tavernier visited those countries in 1616.

In the fifteenth and sixteenth centuries, peers of France had to present bouquets and crowns of roses to the assembled Parliament.

At present there are very important rose plantations in France, Bulgaria, and in the Fayoum in Egypt. In France about ten or twelve thousand roses are grown on two and a half acres. The season is from April to May. Women gather from twenty to twenty-five pounds daily, and obtain from twopence to threepence for two and a half pounds. Each tree will give about a quarter of a pound of roses. The petals are distilled to make rosewater.

Some 12,000 people on the slopes of the Balkans, at Kerzanlik and other places, entirely depend upon their rose plantations. These are on light soil, fully exposed to the sun, at over 1200 feet above the sea. It is interesting to find that the pure mountain air strengthens the perfume, for these Balkan roses are fifty per cent. richer in essences than those of lowland plants.

Another interesting plant much cultivated in the Riviera is the Cassier (*Acacia farnesiana*). It is really a native of India, but was introduced from the West Indies to Europe in 1656. Cannes, Grasse, Antibes, and Nice are the places where it is most cultivated. Its flowers appear from July to November. An old tree may yield as much as twelve to twenty pounds of flowers, worth about five to six francs. But pounds of flowers only yield about a pound of essence, so that it is not surprising that this last is worth £60 the pound.

The cultivation is a little uncertain, for a temperature of three or four degrees below the freezing-point kills the trees.

The pomades made from many of these flowers are produced as follows: A series of trays are covered with fat or grease; the petals are placed on the grease and replaced by fresh petals every twenty-four hours or so; in the end the grease is so saturated with scent that it forms pomade or pomatum.

Thus these half-desert countries are by no means without interest from a botanical point of view. The conditions of life are no doubt hard both for plants and animals. The scent so richly produced depends upon the strong sunlight and pure air. It is very useful, partly because it attracts those useful insects which carry the pollen, but also because such odours are distasteful to grazing animals. The gums, incenses, thorns, and spines are all of great use to the plant in its dangerous struggle for existence with hungry camels and thirsty soil.

When men understood how to irrigate the soil, and before they were foolish enough to cut down the forests which once guarded the mountain springs, these half-deserts were exceedingly prosperous; they were full of vigorous intellectual life, and of strong, hardy, and industrious peoples. Asia Minor, Turkey, Greece, and the Northern Coast of Africa from Morocco to Egypt, were rich and wonderful countries.

But it was not only the destruction of the forests that has ruined them. The curse of Mohammed, the fatalism produced by his religion, and the slavery which is a necessary part thereof, have destroyed the people in mind, body, and spirit. Even in Greece, Algiers, and Cyprus there has been as yet but small recovery.

In the future, not merely these countries, but Northern Nigeria, British East Africa, and South-west Cape Colony, may have as rich a history as Greece, if British brain and energy are helped by the strong muscles of the African.

CHAPTER IX. ON TEA, COFFEE, CHOCOLATE, AND TOBACCO.

ENGLISH TEA-DRINKING – STORY OF OUR TEA – ASSAM COOLIES – MANUFACTURE IN INDIA AND CHINA – CELESTIAL MOISTURE – DANGER OF TEA – THE HERMIT AND HIS INTELLIGENT GOAT – GOVERNMENT, COFFEE AND CAFÉS – CHICORY – CHOCOLATE – AZTECS – KOLA AND ITS CURIOUS EFFECTS – TOBACCO – SIR WALTER RALEIGH – GREAT EMPERORS AND TOBACCO – COULD WE GROW TOBACCO? – STORY OF A SUMATRA CIGAR – DANGER OF YOUNG PEOPLE SMOKING TOBACCO.

ON every day throughout the year English people drink about 600,000 lb. of tea. That is about 270 tons, which would form, when made into the beverage, a lake quite large enough to float a man-of-war! No other civilized nation takes its tea in the reckless way that we do. Yet our fellow-subjects in Australia drink even more than ourselves.

Almost the whole of this tea is grown in British colonies or possessions, manufactured by British subjects, and imported in British ships.

The coolies who work in the tea-gardens of Assam and Ceylon, the Englishman who manages them, the engineers in Glasgow and Newcastle who made the machinery, the shipbuilders, shipowners, and crews, are all fellow-countrymen of those who drink the cup that cheers. Every sixpence in the £8,000,000, which is our yearly account for tea, finds its way into the pockets of our fellow-subjects either at home or abroad.

The women in the Ceylon Tea Gardens bring in their baskets in the evening. These are then weighed as shown and the labourers paid accordingly.

Every one would suppose that a trade like this, which benefits everybody, would be very carefully fostered by Government.

Far from it, for this is one of those articles that are always being attacked by Chancellors of the Exchequer, who seem to have a special ill will against tea.

Indeed, it is so heavily taxed that it is extremely difficult to make a profit on tea-gardens. Elsewhere in this chapter some other very curious facts will be found illustrating the extraordinary habits and methods of the British Government.

The author does not try to explain these facts, but only points them out; a nation that can manage to exist at all when such things are done by its Government is a nation to which one is proud to belong.

The Tea-plant is a native of China and Assam. It is a very handsome shrub resembling a camellia, with dark, glossy, green leaves and beautiful flowers. It is said to have been used in China about 2700 B.C., and the first plantations in India were made with Chinese seed. But a Mr. Bruce reported the presence of an indigenous wild tea in Assam. Three botanists who were sent to investigate the question suggested that this Assam variety was only the Chinese plant run wild, and advised the introduction of Chinese seedlings. This was a very unfortunate mistake, for the wild Assam plant gives much better results.

The jungle is first cut down and cleared away by the native tribes, with the help of elephants. Then at the right season, i.e. after the rains begin, the Indian women and coolies go into the plantations. They carry on their backs a basket supported by a band across the forehead. These women nip off the first two leaves and a bud with their finger and thumb and throw them into the basket over their shoulders. When the basket is full they take it back to the factory, where their gatherings are weighed. The actual manufacture is, in India and Ceylon, all performed by machinery. The tea is first emptied on to trays in a shallow layer: a pound of tea when so spread out covers more than a square yard. These trays are then placed in a room which is heated to a high temperature, for "withering." After six hours it is passed through a machine which "rolls" or gives a twist to the leaves. It is then "fermented" on cement floors, where the tea is covered by strips of moist muslin. It is again rolled and afterwards dried or "fired." The sifting out of the different sorts or blends, and also the packing of the tea in chests, are done by machinery.

That is the Indian system of manufacture, in which there is scarcely any hand-labour.

In China the rolling, and indeed every stage of the process, appears to be done by hand. It is obvious that in the handling, pattings, and rollings of the tea by Chinese coolies, "celestial moisture" may be imparted to it. In spite of this, however, the export of Chinese tea is steadily diminishing. In the old days, the Liverpool "tea clippers," fast and beautiful sailing-ships, raced each other home from China in order to get the first tea upon the market.

Tea is sometimes dangerous, and especially when it is allowed to stew on the fire for hours at a time. Besides *theine*, which is the stimulating, active part of it, and which is a bracing tonic to the nerves, *tannin* is also found therein. When meat is taken with a large amount of *tannin*, the latter acts on the meat exactly as it does on hides in a tanning factory. It forms a substance resembling leather, which taxes the powers of the strongest digestion.

Once upon a time in those fertile mountains of Abyssinia which have never yet been explored by the white man, there was a very holy and pious hermit. He used to live entirely on the milk of a few goats which he carefully tended with his own hands. One morning he noticed that one of these goats showed signs of unusual excitement. It was frisking about, and obviously was exceedingly well pleased with itself.

That was not a usual experience with the holy recluse, who watched the animal carefully. He soon discovered that it was in the habit of grazing on the bright red berries of a very handsome shrub in the hills. The anchorite tasted those fruits and discovered that he also became both pleased with himself and somewhat excited.

His disciples soon discovered a brightness and exhilaration, an unusual "snap," in the good man's sermons, and they watched him and also discovered *Coffee*!

The author refuses to take the responsibility of more than the discovery of the above story. Coffee was, however, introduced into Arabia by the Sheikh Dabhani in 1470. It was taken to Constantinople about 1554, and about a hundred years later coffee-houses and *cafés* were regular and habitual daily resorts in London and Paris.

As usual with stimulants of all kinds, the watchful eye of a moral Government discovered something objectionable in coffee, and Charles II in 1675 imposed heavy taxes, or rather forbade the use of it altogether.

There was in 1718 a coffee-plant in the botanical gardens at Amsterdam, and in that year some of its seeds were sent to Surinam, in Dutch Guiana. Apparently the millions of shrubs in the enormous coffee plantations of the New World are all descended from this particular Amsterdam plant.

This New World coffee is by far the most important supply. Brazil alone exports about £19,000,000 worth of coffee, and that from the New World forms about 82 per cent of the total world's production.

The story of coffee in Ceylon is a tragedy. There happened to be in the jungle a particular fungus (*Hemileia vastatrix*) which got its living on the leaves of wild plants belonging to the coffee order (*Rubiaceæ*) and others. When Arabian coffee was introduced, the fungus began to attack its leaves. The result was the utter ruin of the industry. It is said that about £15,000,000 was lost by this *Hemileia* disease in Ceylon.

The plantations require a great deal of care. The shrubs have to be carefully pruned, and the preparation of the coffee bean is not a very easy matter. It is really the seed of a bright red, fleshy berry. The pulp or flesh has to be removed, and also both a horny skin, the "parchment," and a thin delicate membrane, the "silverskin," in which the seed is enclosed. Coffee is not nearly so much used in Britain as in some other places, and particularly in Holland, for the Dutch drink about twenty-one pounds per head in the year, whilst we in Great Britain only use about three-quarters of a pound.

It is in fact not very easy to make good coffee, and it is absolutely necessary to grind and roast the beans just before using them. Very often also too little coffee is used.

Tinned coffee is often adulterated with either Chicory or Endives, but those are only the two most important impurities, for burnt sugar, biscuits, locust-beans, date-stones, rye, malt, and other substances are ground up and mixed with coffee.

The use of chicory is, however, more or less recognized. It is the roots which are ground up and mixed with it. They contain no *caffeine*, which is the active part of the coffee bean, and are quite harmless. At one time chicory was grown in Essex and other English counties, and was a distinctly profitable crop.

Here again come in the mysterious ways of the British Government. The cultivation of chicory was absolutely forbidden by the Inland Revenue Department; but a considerable amount is still grown in Belgium and is imported to this country. Those who prefer chicory with their coffee have to pay a heavy duty; but the Belgian farmer is allowed and the British farmer is forbidden to take up a paying and profitable industry! The plant is allied to the dandelion. It occasionally occurs in this country as a weed, and is a rather striking plant with bright blue flowers.

Another of these useful productions which also suffers from a heavy duty is Cocoa or Chocolate. There are a great many different plants called Co Co, or by some name very similar to it. The Cocoanut Palm furnishes not only the nuts but the fibre or coir enclosing them, as well as a great many other useful substances. The *cocaine* used by dentists, and which deadens or stupefies the nerves of the teeth, is derived from the leaves of a Peruvian shrub, "Coca" (*Erythroxylan Coca*). These leaves are chewed in the mouth and have very extraordinary effects, especially on the Indian labourers. They are a strong nerve stimulus and take away any feeling of hunger or fatigue. It was by the use of coca leaves that the postmen of the Inca emperors in Peru were enabled to carry messages at the rate of miles a day. Then again the Cocoes of the West Indian Islands is a sort of Yam (*Colocasia antiquorum*). Coco-de-mer is the fruit of a palm common in the Seychelles Islands (*Lodoicea Seychellarum*).

The cocoa which gives the ordinary chocolate and cocoa of the breakfast table is the seed of a tree (*Theobroma cacao*). The name is derived from θεος, god, and βρωμα, food. It may be translated, "That which the gods browse upon."

This plant is one of those which were cultivated by that ancient, powerful, semi-civilized nation, the Aztecs of Mexico. They have almost entirely vanished; at any rate their descendants, if they have any, exercise practically no influence in the world, but they have left us chocolate. They fully appreciated the plant, and even more than we do, for they worshipped it with grateful and superstitious awe.

In their tombs, chocolate flavoured with vanilla was placed, in order to provide the ghost with sufficient sustenance for his or her aerial flight to the Land of the Sun. Columbus brought home some cocoa on his return from his first voyage. The Jesuit fathers in Mexico greatly helped in developing the plantation of cocoa in the days of the Spaniards. At present the largest amount comes from Ecuador, which produces about 50,000,000 pounds weight.

It is a small tree, twenty to thirty feet high, growing in the warm, moist, and sheltered forests of Central and South America. It has a large fruit, within which are the numerous cocoa beans, "nibs," or seeds. The tree does not bear until it is five years old. The fermentation and drying of the beans require some care.

Chocolate is made from the powdered cocoa mixed with sugar and other materials. Chocolate, like tea and coffee, depends for its effect on an extremely powerful drug, *theine* or *caffeine*, of which it contains minute proportions. There are very few other plants known which possess this powerful substance. Amongst these is the Kola nut, which is everywhere regularly employed in West Africa. On the way up to the barracks at Freetown, Sierra Leone, natives were always to be seen seated by the roadside; they sold kola nuts to the soldiers, who were thereby enabled to walk steadily and uprightly past the sentry, and to return his challenge in a clearly articulate voice, although they might previously have been somewhat injudiciously convivial in the town. This kola is one of the very strongest nerve tonics; under its influence men can endure severe physical and mental strain. Like the others, however, a depressing reaction inevitably follows, accompanied by insomnia, headache, and other evil effects.

When one comes to ask, Why do those few plants out of all the vast multitude of the vegetable world possess such extraordinary virtues? it is difficult to find an answer. Possibly some obscure insect or fungus enemy finds *caffeine* poisonous.

Nor can one find any reason for the curious properties developed in the Tobacco leaf by fermentation, except a possible protection to the leaf from the attacks of insects. No doubt the leaf, even in its natural state, would be too strong for them.

Tobacco is a native of Central America. The name *Nicotiana tabacum* is derived (the first) from a certain Jean Nicot, Ambassador to the King of Portugal, and the second from the Haytian name for a pipe.

On Columbus's voyage in 1492 the use of tobacco was noted. The story of Sir Walter Raleigh's servant, who threw a bucket of water over his master when the latter was smoking a pipe, is not supported by much evidence, but it seems to be probable that Sir Walter did smoke his pipe on the way to the scaffold.

At any rate it was cultivated in Europe by the year 1570, and Spenser speaks of the "soveraine Weed, divine Tobacco."

From the first it was detested by all governments and authorities. James I published a very intemperate *Counterblast against Tobacco*. It was prohibited by the Czar of Russia in 1635, and by the King of France. The great Sultan Jehanghir in India, Sultan Amurath II in Turkey, Shah Abbas the Great in Persia, and the Emperor Kang Ching in China, all prohibited the use of tobacco in their respective dominions.

Yet none of these great rulers were able to check its progress. The "Herb of Amiability," or the "Queen Herb of the rude Barbarian" as it is described in Chinese, prevails almost over the whole earth. There is scarcely a people or tribe in existence which does not use it.

But almost everywhere it is either heavily taxed or a Government monopoly; in the latter case it is always exceedingly bad. We ourselves import tobacco worth about £4,500,000 in the year, and pay a heavy duty. The world probably smokes from 1,800,000,000 to 2,000,000,000 pounds of tobacco every year.

The plant is a very pretty one, with large leaves and long pinky or white flowers, which are open and strongly scented at night. It is an annual, and is not at all difficult to cultivate. There is an impression in this country that it is a tropical plant, but by far the greatest amount of our tobacco comes from temperate countries. Large quantities are grown in Germany, in Hungary, and in other parts of Europe. As a matter of fact tobacco was once cultivated in both England and Scotland.

There is evidence to show that in 1832 it was successfully grown in Roxburghshire, where 1000 pounds an acre was obtained. The land was let at about £5 to £6 per acre. Experiments of recent years have also proved very encouraging, and in fact it is difficult to see how any reasonable doubt can exist as to the fact that it would be perfectly easy to grow plenty of that sort of tobacco which we now obtain from Holland and Germany. A prominent Irish statesman has admitted this: "There was no doubt but that tobacco could be grown in Ireland, but whether there are Irishmen patriotic enough to smoke it, *is* very doubtful."

Of course every one knows that the differences in tobacco depend chiefly on the preparation, but the *Constitutional* objection to tobacco, illustrated by the above remark, is the real reason why it is not grown.

Oliver Cromwell sent his troopers to ride down the growing crops. Charles II imposed a penalty of £1600 per acre. Modern statesmen are flippant and unfair.

The reason of course is that a large income is cheaply obtained by taxing imported tobacco. If this were at all interfered with, new taxes, which would certainly be unpopular, would be required.

There is a good deal of interest in the story of the tobacco plantations. Many prisoners of the Civil War in England were sold to Virginia and other places. Even nowadays there is some romance in the history of a cigar. In the Dutch island of Sumatra the jungle is cleared away by the natives under the orders of an English manager. Chinese coolies are then imported. The estate provides each coolie with tools, tea, a barber, and sufficient cash to buy rice, fish, or pork, as well as a little over for his opium, to spend in fireworks, and to propitiate his demons.

The coolie grows the tobacco, which is bought from him and manufactured by the estate. Some of it goes to India, where it is used as the outer wrapper of cigars.

For adulterating tobacco all sorts of leaves are occasionally employed, such as those of the dock, chicory, burdock, foxglove, comfrey, elm, coltsfoot, plantain, beech, cabbage, lettuce (steeped in tar oil), etc., etc.

The substance nicotine is a deadly and dangerous poison. When young people smoke tobacco, it has been quite conclusively proved that they will very probably not reach their full growth, but be miserable weaklings, stunted, half-developed, and below the proper standard of a man.

This is not surprising, if one reflects on the constitution of tobacco smoke. This contains "nicotine, empyreumatic resin, oil, ammonia, carbonic acid, carbonic oxide, hydrocyanic acid, sulphuretted hydrogen, carburetted hydrogen, and paraffin."

CHAPTER X. ON DESERTS.

WHAT ARE DESERTS LIKE?—CAMEL-RIDING—AFTERGLOW—DARWIN IN SOUTH AMERICA—BIG BAD LANDS—PLANTS WHICH TRAIN THEMSELVES TO ENDURE THIRST—CACTUS AND EUPHORBIA—CURIOUS SHAPES—GREY HAIRS—ICEPLANT—ESPARTO GRASS—RETAMA—COLOCYNTH—SUDDEN FLOWERING OF THE KAROO—SHORT-LIVED FLOWERS—COLORADO DESERT—DATE PALMS ON THE NILE—IRRIGATION IN EGYPT—THE CREAKING SAKKIEH—ALEXANDRIA HILLS—THE NILE AND EUPHRATES.

ACROSS the whole of Africa, at its very broadest part, from the dominions of the Emperor of the Sahara at Cape Juby on the Atlantic, and to the very borders of British India, stretches a desert of the most uncompromising character. It is famous in history: the strongest races of man, the great religions of the world, as well as most cultivated plants and domestic animals, have originated in some part of this dreary waste.

One cannot really appreciate deserts unless one has really seen them. But it is necessary to try to describe what they are like.

Sometimes the desert is a wilderness of broken, stony hills covered by angular pieces of shivered rock. In other places the soil is hard, and is everywhere covered by pebbles or shingle. Often it is a mere waste of sand blown into downs and hillocks which look sometimes like the sand dunes by the coast, and elsewhere like the waves of the sea.

One finds valleys in the desert quite like ordinary ones in shape, but instead of water there is only sand in sweeping curves and hollows, like the snow-wreaths and drifts in a highland glen.

Rocks stand out of this, but their projecting faces are polished smooth and glittering or deeply cut by the flinty particles scraping over them continually in storms and hurricanes.

The traveller on camel-back, where his waist has to act as a sort of universal joint giving to every unexpected jolt and wrench of his rough-paced mount, suffers from the heat, for nowhere else in the world are there such high temperatures. He suffers from thirst, and still more from the dust which fills eyes, mouth, nostrils, and ears.

Yet the dry pure air is most exhilarating.

In the evening there is a feast for the eyes in the glorious *afterglow* when the sun has just set. The light from below the horizon produces an ever-changing, indescribable play of colour from violet to salmon pink and through the most delicate shades of yellow, blue, and rose, until everything fades and there reigns only the mysterious silence of the beautiful starlit night.

No wonder the air is dry and pure, for rain only falls on perhaps eight days in the year in some places (Ghardiaia).

Yet plants manage to exist even where there is only about seven inches of rain annually.

But this seems still more extraordinary if one remembers that sand may be almost glowing hot during the day, whilst in winter it may be, at night, cooled below the freezing-point.

Yet a desert absolutely bare of plants is an exceedingly rare phenomenon. Such do occur. Darwin speaks of "an undulating country, a complete and utter desert." This is not very far from Iquique in South America. "The road was strewed with the bones and dried skins of the many beasts of burden which had perished upon it from fatigue. Excepting the *Vultur aura,* which preys on the carcases, I saw neither bird, quadruped, reptile, nor insect. On the coast mountains, at the height of about 2000 feet, where, during this season, the clouds generally hang, a very few cacti were growing in the clefts of rock; and the loose sand was strewed over with a lichen which lies on the surface quite unattached. ... In some parts it was in sufficient quantity to tinge the sand, as seen from a distance, of a pale yellowish colour. Farther inland, during the whole ride of fourteen leagues, I saw only one other vegetable production, and that was a most minute yellow lichen, growing on the bones of the dead mules."

Rydberg, speaking of the Big Bad Lands in South Dakota, says that there are in some places great stretches of land consisting of cañons separated by small ridges, in which not a speck of green is visible over several sections. (A section is more than a square mile.)

But though Aden looks exactly like "a barrack stove that no one's lit for years and years," plants grow there. Even in Egypt, when one has left the Nile inundation limit, a botanical eye very seldom fails to detect plants of one sort or another even in a dangerous and thorough-going desert.

Plants are almost as hardy as men; they can adapt themselves to almost any climate.

In some curious and inexplicable way the very dangers of the climate seem to produce automatically a means of resisting it. The chief peril, of course, is a loss of the precious water through the leaves. When the skin or epidermis of a plant is being formed, the walls of its cells are laid down, layer by layer, one inside the other, by the secretion of the living matter inside. In a dry desert the loss of water by evaporation will be so rapid that these layers of cell-wall are much thicker than in ordinary plants. The very fact that they are thicker and less penetrable tends to prevent any further loss of water.

So that plants in a dry climate have the power of altering themselves to resist its dangers.

Another author found that, in Scandinavia, plants of the same species can acclimatize themselves if necessary. Sheep's Sorrel which had grown on dry, droughty gravel banks only lost 10 per cent. of its water in the first two days, when it was artificially dried. Other Sheep's Sorrels, which had been luxuriating in meadows where they had no lack of moisture, lost no less than one third (33 per cent.) of their water when dried in the same way.

That is interesting, because very likely our readers might in crossing a desert be perishing of thirst when a Bedouin Arab would be perfectly happy. The plants have learnt to do without water just in the same way as the Arab has done.

Of the many interesting desert plants, the Succulents, Cacti, Euphorbias, and others of the same extraordinary, fleshy, dropsical appearance, come first.

When a Cereus plant (one of the American Giant Cacti) was dried, it did not lose the whole of its water for 576 days. That is probably the longest time "between drinks" on record. A Houseleek (*Sempervivum*), which has to grow on dry rocks where it has no water for days together, remained quite fresh for 165 days.

This plant was about feet high. The darker part on the tallest branch is the dark red flower of the parasitic horanthus. The thorns covering the branches are quite distinct.

There are several reasons why these plants took so long to dry up. To begin with, they have inside their stems and leaves certain substances which hold water and delay its escape. Moreover their extraordinary shapes are of very great assistance. They prefer globular, round, circular, pear-shaped, or cylindrical forms.

Suppose you were to cut such a round mass into thin slices and lay them out flat, it is quite clear that they would cover a much greater surface. Thin leaves also, if squashed up into a round ball, would have a very much smaller surface.

The water can only escape from the surface exposed, so that these condensed round balls and fleshy columns have far less water-losing surface than ordinary leaves.

As a matter of fact, it was found by calculation that the surface of an Echinocactus was 300 times less for the same amount of stuff as that of an Aristolochia leaf. If the actual loss of water from the Echinocactus, as found by experiment, was reckoned as one unit per square inch, then the amount of water lost from a square inch of the Aristolochia was no less than 5000 units!

This shows that these odd, outrageous shapes of Prickly Pears, Cacti, and other succulents are an extraordinary help to them. We have already pointed out in a previous chapter how necessary their spines and prickles are if they must resist rats, mice, camels, and other enemies.

What we may call the "hedgehog" type of plant is also very common in desert countries. There are many woody little, much branched, twiggy shrublets, which bristle all over with thorns and spines. They are not at all fleshy, but do with the least conceivable amount of water.

Another striking characteristic of the desert flora is noticed by every one. Almost every plant is clothed either in white cottonwool, like the Lammie's Lug of our gardens, or else in grey hairs. The general tint of the landscape is not green, but it is rather the colour of the soil silvered over by these grey-haired plants.

The reason of this is, of course, quite easy to understand. We put on a thick overcoat if we are going to walk in a Scotch mist, to keep out the moisture. These plants cover themselves with hairs or cottonwool to keep the moisture inside. It does not escape easily through the woolly hairs on the skin.

One very strange plant should be noticed here. This is the Iceplant (*Mesembryanthemum cristallinum*). Every part of it is covered with little glittering swellings which shine in the sun like minute ice crystals. The swellings contain a store of water, or rather of colourless sap, which makes it able to exist in dry places. Dr. Ludwig says that a torn-off branch remained quite fresh for months on his study table. It is probable that these peculiar pearl or ice-like swellings also focus the sunlight, acting like lenses, upon the inner part of the leaf, but that is not as yet fully understood.

There are two grasses, growing in the desert, which are of some value; both are called Esparto or Halfa. They are very dry, woody, or rather wiry grasses, especially common in Algeria, Tripoli, and also found in Spain. One of them, *Stipa tenacissima*, grows in rocky soil in Morocco, Algeria, and Tripoli. The Arabs search for it in the hills, and dig it up by the roots; they then load their camels with the grass and bring it to the ports whence it is sent to London or other places. A very good and durable paper is made from it, and ropes, mats, and even shoes are also produced from the fibre. Part of the "esparto" is, however, furnished by another grass (*Lygeum sparteum*). The natives sometimes tie a knot in a halfa leaf, which, according to them, cures a strain of the back. The Stipa is also used as fodder, but it is not nutritious and is indeed sometimes dangerous. In one year Britain imported 187,000 tons of esparto, worth nearly £800,000. The yield is said to be about ten tons per acre.

Another very interesting plant at Tripoli and in the North African Desert generally, is a sort of broom, the Retama (*Retama Raetam*). It is not very unlike the common broom, but has long, leafless, whip-like branches covered by bright pink-and-white flowers. It can often be seen half submerged in waves of sand, and struggling nevertheless to hold its own. As it has no leaves its loss of water is very much kept down. This is the Juniper of the Bible, and it is still used for making coals.

The length of the roots is very great in most of the broom-like, "hedgehog," and other plants. A quite small plant not more than six or eight inches high will have a root as thick as one's thumb. Even at a depth of four or five feet below the surface its root will be as thick as the little finger, so that the root-length is at least twenty times the height of the visible part above ground. These thirsty roots explore the ground in every direction, and go very deep downwards in their search for water.

Another very interesting plant in the Egyptian Desert is *Citrullus Colocynth,* from which the drug colocynth is prepared. The great round yellow-green fruit and finely divided bright green leaves may be seen lying on the sand. It remains green all the summer, but appears not to have any particular protection against loss of water. It is always supplied by its roots with underground water. If a stem is cut through it withers away in a few minutes. This is found also in Asia Minor, Greece, and Spain. The pulp of the fruit contains a strong medicinal substance; it is a drastic purgative, and in overdoses is an irritant poison. This was probably the Wild Vine or gourd which the young prophet gathered, and which produced "death in the pot." He probably mistook it for a water melon. It is still plentiful near Gilgal (2 Kings xiv. 38-41).

Below the surface of the earth, of course, there is not nearly the same dryness or danger of losing water, so that there are often a great number of bulbs, tubers, and the like hidden in the soil. There they wait patiently, sometimes for a whole year or even for a longer period. So soon as a shower of rain falls they start to life, push out their leaves, and live at very high pressure for a few days. After a shower of rain, the Karoo in South Africa, for instance, is an extraordinarily beautiful country. There are bulbous Pelargoniums, a very curious leafless cucurbitaceous plant (*Acanthosicyos*), hundreds and thousands of Lilies, Irids, and Amaryllids. A single scarlet flower of a Brunsvigia can be seen more than a mile away!

These tender and delicate, exquisitely beautiful bulbs flourish amongst the succulent Euphorbias and Mesembryanthemums, between the hedgehog-like thorny plants and the woody little densely-branched mats of the permanent flora. The rain stimulates even these last to put out green leaves and flowers, but their time comes later on, when by the return of the usual drought every leaf and flower and the fruit of every bulb has been shrivelled up, turned into powder, and scattered in dust by the wind.

Then the Karoo becomes unlovely, desolate, and barren-looking, with only its inconspicuous permanent plants visible.

The above description applies to bulbs and perennial plants with underground stores of food. Yet these are by no means the only plants which manage to exist in the Egyptian and Arabian desert. After a shower of rain a whole crowd of tiny annuals suddenly develop from seed; they come into full flower and have set their seed before they are killed off by a return of the desert conditions, when the effects of the rain have died away. These plants are not really desert plants at all, for they only grow during the short time that it is not a desert. They are like the Ephemerid insects which live for a summer day only.

Nor is it only in Egypt that we find such ephemerals. Mr. Coville found them in the Colorado desert in North America. The plants are quite different, but similar conditions have brought about an entirely similar mode of life on the other side of the globe! In Colorado they seem to be much influenced by the quantity of rain. Mr. Orcutt, after the great rain of February, 1891, found plants of Amaranthus (allied to our Love-Lies-Bleeding), which were ten feet in height, but in 1892 he found specimens of the same in the same place only nine inches high, though they were perfect plants and in full flower; in this last year there was only the usual very scanty rainfall.

It is, however, in deserts when man has set to work and supplies water and strenuous labour, that the most wonderful results appear. The whole of lower Egypt, Babylon, Nineveh, Damascus, Baghdad, Palmyra, and other historic cities, show what the desert can be made to produce.

As one slowly steams up the Nile from Philae or Shellal towards Wady Halfa, there are places where the brown, regular layers of the Nubian Sandstone form cliffs which advance almost to the water's edge. Yet there is a narrow strip of green which fringes the water.

It is upon the actual bank itself, which is a gentle slope of ten to fifteen feet, that Lupines, Lubia beans, and other plants are regularly cultivated. This narrow green ribbon remains almost always on each bank. Where the cliffs recede, one notices a line of tall, graceful date palms, mixed occasionally with the branched Dôm palm (the nut of which yields vegetable ivory). Tamarisks, conspicuous for their confused, silvery-green foliage, can be noticed here and there. The Acacias are common enough, and sometimes one of them is used as a hedge. It is a spreading, intricately-branched little shrub, with very white branches and stout curved thorns.

If one lands and strolls along the banks below the palm trees or amongst plantations of barley, wheat, or lentils, one sees the native women in their dark green robes gathering fruits or digging. Goats and donkeys are tethered here and there. There are sure to be castor-oil bushes. Small but neat pigeons, with a chestnut-coloured breast and bluish-banded tails are perching on the palms or acacias, and utter their weak little coo. The air is suffering from the horrible creaking and groaning of a "sakkieh" water-wheel. This is made entirely of acacia wood, and is watering the plantations. Sometimes it seems like a crying child, then, perhaps, one is reminded of the bagpipes, but its most marked peculiarity is the wearisome iteration. It never stops. One of them is said to supply about 1-1/2 acres daily at a cost of seven shillings per diem. Exactly the same instrument can be seen pictured on the monuments of Egypt 4000 to 5000 years ago. The "shadouf" is of still older date. This is a long pole bearing at one end a pot or paraffin tin and balanced by a mass of dried mud or a stone. All day long a man can be seen scooping up the coffee-coloured water of the Nile and pouring it on the land for the magnificent sum of one piastre a day.

Where not irrigated, the soil is dry and parched and can only carry a few miserable little thorny bushes. The entire absence of grass on the brick-like soil has a very strange effect to English eyes.

The Date Palm, however, requires a little respectful consideration. If one enters a thick grove and looks upwards, the idea of Egyptian architecture as distinguished from Gothic and others is at once visible. It has quite the same effect as the great hall of columns near Luxor. The numerous stems ending in the crown at the top where the leaves spring off was quite clearly in the minds of the architect at Karnak and other temples. It goes on bearing its fruits for some two hundred years, and begins to yield when only seven years old. It revels in a hot, dry climate with its roots in water, and seems to require scarcely any care in cultivation. Yet during the first few years of its life it is necessary to water the seedling. A single tree may give eight to ten bunches of dates worth about six shillings. Generally it is reproduced by the suckers which spring out from the base of the tree.

Dates make a very excellent food, not merely pleasant but both wholesome and nutritious. Sometimes toddy is made by fermenting the sap, but this is a very wasteful process, as it is apt to kill the tree.

The stones are often ground up to make food for camels. The feathery leaves are exceedingly graceful. When quite young they are not divided, but they split down to the main stalk along the folds, so that a full-grown leaf affords but little hold to the wind.

In some parts of Egypt, as for instance at Mariout, which is some fifteen miles from Alexandria, the wild flowers are probably more beautiful than anywhere else in the world. Amongst the corn and barley, which can be there grown without irrigation, masses of scarlet Poppies and Ranunculus are mingled with golden-yellow Composites, bright purple Asphodels, and hundreds of other Eastern flowers. The result is a rich feast of colour indescribable and satisfying to the soul.

So that these deserts under the hand of man rejoice and blossom as the rose.

Why is it that, as Disraeli has pointed out, civilization, culture, science, and religion had their origins in the desert? The answer is not difficult to see: for there is a dry, healthy climate; the severe strain of a long day's journey is varied by enforced leisure, when, resting at his tent-door, the Arab is irresistibly compelled to study the stars and to contemplate the infinite beauty of the night. It seems also to have been in the desert of the old world that man first learnt to cultivate the soil. In fact, it was only by irrigation on great tracts of alluvium, such as were furnished by the Nile and Euphrates, that the enormous populations of Egypt, Babylon, Nineveh, and the other great monarchies could be maintained. So that city life on a big scale first developed there.

CHAPTER XI. THE STORY OF THE FIELDS.

WHAT WAS ANCIENT BRITAIN?—MARSHES AND BITTERN—OAK FOREST—PINES—SAVAGE COUNTRY—CORNFIELD—FIRE—ICE—FOREST—WORMS—PALEOLITHIC FAMILY—THE FIRST FARMERS—ALFRED THE GREAT'S FIRST GOVERNMENT AGRICULTURAL LEAFLET—DR. JOHNSON—PRINCE CHARLIE'S TIME—MISERY OF OUR FOREFATHERS—OATMEAL, MILK, AND CABBAGES—PATRICK MILLER—TENNYSON'S *NORTHERN FARMER*—FLOURISHING DAYS OF 1830 TO 1870—DERELICT FARMHOUSES AND ABANDONED CROFTS—WHERE HAVE THE PEOPLE GONE?—WILL THEY COME BACK?

WHEN the eyes of man first beheld Britain, what sort of country was this of ours? It is very interesting to try to imagine what it was like, but of course it is a very difficult task. Still it is worth the attempt, for we ought to know something of what has been done by our forefathers.

Where the great rivers Thames, Humber, Tyne, Forth, Clyde, Mersey, and Severn, approached the seashore they lost themselves in wildernesses of desolate, dreary fenlands. Here a small scrubby wood of willow, birch, and alder; there a miles-wide stretch of reeds and undrained marsh intersected by sluggish, lazy rivers, or varied by stagnant pools. The bittern boomed in those marshes. Herons, geese, swans, ducks, and aquatic birds of all sorts found what is now Chelsea a paradise, only disturbed by the eagle, harrier-hawk, vulture, and the like.

Neither at the mouth nor even much higher up in its valley-course, was a river a steady stream in a defined bed. Such beds as it had were probably four or five times their present width; they would be quite irregular, meandering about, changing at every flood, full of islands, loops, backwaters, and continually interrupted by snags of trees.

The rolling hills of the lowlands would be an almost unbroken forest of oak, except where perhaps level land and the absence of drainage produced a marsh or horrible peat-moss. But when we say forest, we do not mean a glorified Richmond Park.

In good soil there might indeed be tall and magnificent trees. But it would be quite impossible to see them! The giants of the forest would be concealed in an inextricable tangle of young trees, brushwood, fallen logs, creepers, and undergrowth. Where the soil was sandy or stony, it might be a scrub rather than a forest, of gnarled, twisted, and stunted oaks, or possibly thickets of sloe, birch, rowan, hawthorn, brambles, and briers.

Every stream would be "wild water" leaping down waterfalls and cutting out irregular, little woody ravines. Here and there boulders and escarpments of rock would break through the forest soil, which would be mossy, thick with undergrowth, and entangled with rotting fallen trunks and branches, crossing at every conceivable angle. The higher hills were covered by a dreary, sombre pine forest. It was of a monotonous, desolate character. Greenish-grey tufts of Old Man's Beard lichen hung from the branches. The ground, treacherous, and broken by boulders, peaty hollows, and dead logs, would be shrouded in a soft, thick cushion of feathery Mosses, with Blaeberry, Ferns, Trientalis, Linnea, Dwarf Cornel, and other rare plants. Through it descended raging and destructive torrents which here might be checked and foamed over dead logs, whilst in another place they cut out bare earth-escarpments or started new waterfalls which ate back into the hills behind.

At the summit of the higher hills, bare rock crags projected out of occasional alpine grassy slopes, or irregular terraces, ravines, and gullies. Below, these alpine ravines ended in a peat-moss, which scattered, dwarfed, distorted, and miserable-looking Scotch Firs and Birches painfully endeavoured to colonize. Here and there on very steep hillsides, wiry, tussocky grass might be growing instead of forest or peat.

A horrible, forbidding, and desolate land, where Deer, Irish elk, bison, bear, wolf, boar, wolverine, badger, and fox, alone enjoyed themselves.

Now consider our country to-day. Mark the "trim little fields"; "that hedge there must have been clipt about eighty years"; "The lifting day showed the stucco villas on the green and the awful orderliness of England—line upon line, wall upon wall, solid stone dock and monolithic pier." The road, carefully macadamized, sweeps on correct and straight or gracefully curving from neat village to countrytown. In the heart of the country the roadsides are scraped bare to produce that hideous tidiness which is dear to the soul of the County Council roadman. That is if an individual whose life is spent in stubbing up roses, briers, and every visible wild flower, can possibly possess a soul! Those fields without a rock, or even a projecting stone, have been drained, dug over, and levelled with the greatest possible care. The very rivers have been straightened and embanked; the rows of pollarded willows have been planted; they may, when in flood, overflow, but the results are very soon no longer visible. Even on the moors and in the depths of the Highlands, black-faced sheep, draining, and the regular burning of the heather, have quite transformed our country.

The original woods have long since vanished: those which now exist are mostly quite artificial plantations, and the very trees are often strangers to Britain.

The story of the Herculean labour by which our country, once as wild and as savage as its early inhabitants the Icenians and Catieuchlanians (and probably with lineaments as barbarous as those of the Coritanian and Trinobant), has been changed to peaceful, fertile meadowlands or tidy arable, is one long romance. To tell it properly would require a book to itself. In this chapter we shall only try to sketch what may have happened on one particular cornfield which exists on the trap-rocks of Kilbarchan, near Glasgow. The reader must bear in mind that even this is a very ambitious attempt! It is an exceedingly difficult undertaking.

The subsoil in this particular cornfield (on Pennell Brae) lies upon the trap-rock formed by one of those gigantic lava-flows which cover that part of Renfrewshire. The whole district at that time must have been exactly like Vesuvius during the late eruption. Its scenery in this early miocene period consisted of glowing molten rock, accompanied by flames of fire, electrical storms, clouds of gas, dust, ashes, and superheated steam

The same landscape at the present time. Notice how the outline of the hills has been softened and its shape rounded. The forest has almost vanished and the river is bridged and confined within definite banks; in fact, only the ravine remains much the same.

Every plant and every animal must have been exterminated. That was unfortunate, for, at that time, Pines, Oaks, Guelder Rose, Willows, as well as Sequoias allied to the Mammoth tree and Sassafras, may have lived in

Scotland along with tapirs, opossums, marsupials, and other extraordinary beasts.

When the lava cooled and became trap-rock, it was at once attacked by frost, by wind, and by rain. Then by a very slow process of colonization, vegetation slowly and gradually crept over the trap-rock and rich mould and plant remains accumulated. At a much later date, there was another wholesale destruction. This time, it was the great Ice Sheet coming down from the Highland hills. Probably it drove heavily over the top of Pennell Brae and worked up into fine mud and powder every vestige of the miocene vegetation.

The very rocks themselves would be scratched, polished, and rounded off. When the glaciers melted away and left the surface free, it would consist of these rounded rocks alternating with clay-filled hollows. The trap-rock below would be covered by a subsoil due to particles of trap, of Highland and other mud, with remains of the miocene vegetation. Upon this surface, frost, wind, sunshine, and rain would again begin to perform their work.

But the subsoil, thus wonderfully formed by fire in the miocene, by frost in the glacial, and by weather in our own geological period, very soon felt the protecting and sheltering effect of a plant-covering.

First a green herb rooted itself every here and there amidst the desolate boulder-clay or perhaps in a crevice where good earth had accumulated. Then the scattered colonists began to form groups; soon patches of green moss united them. Then a continuous green carpet could be traced over a few yards here and perhaps on a few feet somewhere else. But when things had got as far as this, progress became much more rapid, and soon the whole site of the future cornfield was covered over by a continuous green carpet. Only, every here and there, hard stones and uncompromising trap-rocks remained still protruding from the green covering.

In another chapter this first covering of the soil will be described at length.

So far it has been *subsoil* and *underlying rock*, but now the roots begin to disintegrate and work up the subsoil; the earthworm has his chance, and forms true *soil*. On this particular hillside, the water would drain away and there would be no danger of mosses strangling and choking the Blaeberry and the Heather. The worm flourished and multiplied, and the soil became rich and black. Here and there a Sloe or a Rowan, or Poplar, or perhaps Alder and Birch, began to appear. In certain places Whins and Brooms, Brambles and Briers, diversified the hillside. Then a few Scotch firs began to push their way up, through the thickets. At first they were very small and stunted, but as each one formed a dense, deep-going mass of hardy roots, they were able to investigate the riches of the subsoil. Every year the amount of leaf-mould above increased, until the original moss-covering was utterly destroyed and a pine forest (see Chap. XXVIII.) occupied Pennell Brae.

About this time, a paleolithic family may have encamped on the side of the cliff near a little stream which can still be traced. The camp was only a few sticks and branches, with a skin or two for shelter from the north wind. The women lopped down fir branches for firewood, and cut up the young trees. The children set fire to the shrubs on dry days and paths ran here and there through the forest. This would be about 198,000 B.C.

Every year meant a further very gradual, slow destruction of the pine forest.

About 60,000 B.C., our paleolithic hunters with chipped-stone weapons would be obliged to travel further to the north. New savages with round heads and polished-stone weapons would make life in Renfrewshire too uncertain and too diversified by massacres. These last possessed seed corn, a few fruit trees as well as goats, cattle, and perhaps a few hardy, shaggy ponies. At first these settlers would be obliged to live in a lake dwelling, say in Linwood Moss, which is close at hand. They would then drive their cattle over the surrounding district, and camp in slightly-built villages. Near at hand, probably on the hill, they would build a (round) camp or fort, where they could fly for safety in the continual fights and invasions of the period.

Sooner or later a village would be built near Pennell Brae. One summer day the villagers attacked the wood that covered it; they cut down all the small brushwood and hacked through the bark of every big tree. After a few weeks, when the trees were dead, the wood was set on fire. Then a rough fold made of rude wattle and daub was formed, and every night the cattle and sheep were driven in.

After three or four years, this fold would be ploughed up by exceedingly rude instruments. Barley or certain kinds of wheat would be grown year after year until the crop was not worth gathering. When that happened, another fold would be ploughed up. Probably the whole of Pennell Brae went through this rude sort of agricultural treatment at one time or another. At the same time goats, cattle, and the demand for firewood, obtained in the most reckless and wasteful manner, would have very seriously interfered with the forest.

Although no doubt great changes for the better were introduced, the spearmen of Wallace of Elderslie close by had their "infield" land, which was practically the sheepfold as above described, and their "outfield" or grazing commons. Even down to 1745 the above system was practised (see below).

But when men's minds were stirred up and invigorated by the great Revolution of 1788-1820, all sorts of new agricultural discoveries were made. Yet the cornfield on Pennell Brae was probably not drained or enclosed by stone walls and hedges until 1830 to 1840! About 1870, it was more profitable to its owner than it has ever been since, though even now it forms part of our British farmlands which yield, on the whole, a larger amount of oats per acre than those of any other country in the world (except possibly Denmark).

Let us however look a little closer into the long, long period during which the "fire and stone-axe methods" of farming prevailed. Before the Romans landed there seem to have been no towns. There was but little cultivation, for the Britons wore skins and lived chiefly on milk and flesh.

In the time of King Alfred, the increase of population made it necessary to take more trouble about farming, so we find a description of what the good farmer ought to do. We might call this the very first Government leaflet, and it has led to the Agricultural Leaflets published by the Board of Agriculture for Great Britain and Ireland.

"Sethe wille wyrcan wastbaere lond ateo hin of tham acre aefest sona fearn and thornas and figrsas swasame weods."

He was to clear off fern, bracken, thorns, sloe, hawthorn, bramble, whin, and weeds. The names of the months give some idea of Anglo-Saxon methods of farming. May was *Thrimylce,* because the cows might then be milked thrice a day. August was *Weodmonath* (weed-month), November *Blotmonath,* or blood-month, because the cattle were then killed to supply salt beef for winter time.

Very much later in history, after our English friends had laid waste and depopulated Scotland, so that woods sprang up again everywhere, and again long after that time when the gradual increase of population had again utterly destroyed those woods, a certain Dr. Johnson travelled from Carlisle to Edinburgh. This gentleman declared that he saw no tree between those places. This statement must not be taken too literally, for he had written a dictionary and considered himself not merely the *Times* but an *Encyclopædia Britannica* as well.

The Earl of Dundonald (in 1795) thus describes the agriculture of 1745 (Prince Charlie's days): "The outfield land never receives any manure. After taking from it two or three crops of grain it is left in the state it was in at reaping the last crop, without sowing thereon grass-seeds for the protection of any sort of herbage. During the first two or three years a sufficiency of grass to maintain a couple of rabbits per acre is scarcely produced. In the course of some years it acquires a sward, and after having been depastured for some years more, it is again submitted to the same barbarous system of husbandry" (that is used as a fold and then ploughed up). In the same year (1745) in Meigle parish, the land was never allowed to lie fallow: neither pease, grass, turnips, nor potatoes were raised. No cattle were fattened. A little grain (oats or barley) was exported. In or thereabouts, there was only one cart in the parish of Keithhall. Everything was carried about on ponies' backs, as is the case nowadays in the most unsettled parts of Canada. The country in places was almost impassable. Bridges did not exist, and the roads were mere tracks. In Rannoch the tenants had no beds, but lay on the ground on couches of heather or fern.

These houses were built of wattle and daub, and so low that people had to crawl in on hands and feet and could not stand upright.

"In the best times that class of people seldom could indulge in animal food, and they were in use to support themselves in part with the blood taken from their cattle at different periods, made into puddings or bread with a mixture of oatmeal. Their common diet was either oatmeal, barley, or bear, cleared of the husks in a stone trough by a wooden mallet, and boiled with milk; coleworts or greens also contributed much to their subsistence, and cabbages when boiled and mashed with a little oatmeal." Potatoes were introduced in Dumfriesshire some time after 1750, and the use of lime as manure at about the same time. Even in 1775 the roads were such that no kind of loaded carriages could pass without the greatest difficulty.

There is a most fascinating account in Dr. Singer's work of a strong man's difficulties in starting reasonable agriculture in Dumfriesshire about the year 1785. This was Patrick Miller, of Dalswinton. (It was on Dalswinton Loch that he tried the very first steamboat.) "When I went to view my purchase, I was so much disgusted for eight or ten days that I then meant never to return to this county. A trivial accident set me to work, and I have in a great manner resided here ever since.... I have now gone over all of this estate, and this I have done without the aid of a tenant.... I need not inform you that the first steps in improvement are draining when necessary, inclosing sufficiently, removing stones, roots, rubbish of every kind, and liming.... These operations cost me, I reckon, about £11 per acre upon an average; and I lay my account with being repaid all my expenses by the first three crops, but at any rate by the fourth. When the land which I make arable will give at least (if brought from a state of nature) twenty times the rent when I began to improve it."

Major-General Dirom, of Mount Annan, writing from that place in, says that all over Scotland for about thirty years (from 1780-1810) he has seen "cultivation extending from the valleys to the hills, commons inclosed, wastes planted, and heaths everywhere giving way to corn: ... extension of towns and villages, by new lines of excellent roads, magnificent bridges and inland navigation ... our rapidly increasing population, by our now exporting great quantities of grain from parts of Scotland into which it was formerly imported, and by the superior comfort and abundance which appear in the domestic economy of the inhabitants." If you read any newspaper of to-day published in Canada or in the Argentine Republic, you find exactly the same process at work, and the same enthusiasm about it. Even in 1840-1850 all these improvements were still vigorously going on.

Look at Tennyson's *Northern Farmer* (old style): —

"'An I a stubb'd Thurnaby waäste.
Dobbut looök at the waäste, theer warnt no feeäd for a cow,
Nowt at all but bracken an fuzz, an looök at it now.
Warnt worth nowt a haäcre and now there's lots o feeäd,
Four scoor yows upon it an some on it down i seeäd."

Even in his days, the good farmer was following King Alfred's directions. About 1830-1850 most of the land was in good bearing, and the roads were sufficiently good to admit of the stage-coach with four horses. But they after all lasted but a very short time before the railways again entirely altered the conditions of country life.

As we have seen, rents were in places, five times as large in 1820-1830 as they had ever been previously. Therefore it was that about this time the gentlemen's houses were in many places rebuilt on a more magnificent scale. Then also were begun those circles and strips, or belts of plantation, which are now conspicuous features of the Scotch lowlands. An enormous majority of these plantations are not more than eighty years old. I think avenues were planted in the seventeenth and eighteenth centuries. The fashion about 1820 was to destroy them as unnatural, at least in England. Unfortunately no respect was paid to the economic practice of forestry, with very unfortunate results for the proprietor.

The rest of this chapter is necessarily unpleasant and distressing reading, but it is necessary if we are to understand the romance of the fields. As one wanders over the grassy pastures of Southern Scotland, where the black-faced sheep foolishly start away, and where one's ears are irritated by the scolding complaints of the curlew or whaup, it is no rare accident to find a few broken-down walls, a clump of nettles, and badly grown ash trees. That was once a farm steading, where a healthy troop of children used to play together after walking three or more miles barefoot to school. The ash trees were planted at every farm "toon," for the Scottish spear was a very necessary weapon until recent times. Often also, upon some monotonous grouse moor, one sees the ridges that betoken a little croft where a cottager lived.

In one parish (Troqueer) over seventy country cottages have been abandoned during the lifetime of a middle-aged person.

Many families, of which the laird was often the best farmer in the district and his own factor, have disappeared. The fine houses, with their parks and shootings, are let to strangers, who come for a few weeks or months, and then leave it in charge of a caretaker. Before this recent development, the "family" lived all the year round upon the land; they spent their income chiefly in wages to the country people. Where once forty or fifty people were employed all the year, there are now but three or four. The big house with shuttered windows and weed-grown walks, is a distressing and saddening spectacle.

Of course such changes must occur. The farmer's and the cottar's children are now carrying out in Canada, Australia, or the United States, what was done in Scotland from 1780-1830. India, South Africa, and China have been developed by the brains and hold the graves of many of the laird's sons.

Yet this poor old country, abandoned of her children, shows signs of revival. Both the poor and the rich are beginning to find out that a country life is healthier, quite as interesting, and sometimes quite as profitable as the overcrowded city with its manufactories, mills, and offices. All new countries are beginning to fill up, and there is some hope that a new and vigorous development of farming may make the countryside once more vigorous, prosperous, and full of healthy children.

CHAPTER XII. ON PLANTS WHICH ADD TO CONTINENTS.

LAKE ARAL AND LAKE TSCHAD—MANGROVE SWAMPS OF WEST AFRICA—NEW MUDBANKS COLONIZED—FISH, OYSTERS, BIRDS, AND MOSQUITOES—GRASPING ROOTS AND SEEDLINGS—EXTENT OF MANGROVES—TOURADONS OF THE RHONE—SEA-MEADOWS OF BRITAIN—FLOATING POLLEN—REEDS AND SEDGES OF ESTUARINE MEADOWS—STORMS—PLANTS ON SHIPS' HULLS—KELPS AND TANGLES IN STORMS—ARE SEAWEEDS USELESS?—FISH.

THE way in which the savage, rugged, inhospitable Britain of the Ice Age changed into our familiar peaceful country formed the subject of the last chapter.

But plants do far more than cover the earth and render it fertile, for some of them assist in winning new land from the sea or from freshwater lakes. The Sea of Aral, for instance, or Lake Tschad are rapidly becoming choked up by reeds and other vegetation. Blown sand from the deserts around is caught and intercepted by these reeds, so that fertile pastures are gradually forming in what used to be the open water of a deepish lake.

By far the most extraordinary of all these plants which form new land are the Mangroves.

They are only found in the tropics or sub-tropical regions, and are always along the sea-coast. It is where a river ends in a delta, dividing into intricate and confused irregularly winding creeks, that the mangroves are especially luxuriant.

Such a river will have probably flowed through hundreds of miles of the most exuberant tropical forest, where growth is never checked by the cold grasp of winter.

Its waters are yellowish brown or *café au lait* coloured, because they are full of mud and of decaying vegetation, with dead leaves and decaying branches floating on the surface. So full are such rivers of decaying material that they have a distinct and unmistakable smell, which has been compared to "crushed marigolds."

So soon as the muddy water reaches the sea, most of its mud is deposited and forms great banks and shoals of shifting odoriferous slime, which confuses and interferes with the discharging mouths of the river.

It is upon these changing, horrible-smelling banks of bottomless slime that the Mangrove is especially intended to develop.

If one takes a canoe in such a delta and paddles inwards on the incoming tide, a dense forest of glossy-green mangroves will be found to cover the whole coast-line, and also to extend far inland by the winding creeks, lagoons, and river channels.

The whole theory of the mangroves becomes clearly revealed as soon as the water begins to sink at low tide. First one notices that the stem of every mangrove ends below, not in a single trunk, but in an enormous number of arched, stilt-like supporting roots. Not only the stem but the branches also give off descending roots, which branch into four or five grasping arched fingers as soon as they get near the water. When they reach the mud, these fingers grow down into it and form a new supporting root to the tree. It is very difficult to give any idea of the extraordinary appearance of these mangrove roots.

Imagine an orchard of very old apple trees in winter, and suppose that one were to cut off every tree and plant it upside down in black mud, and also to crowd them so closely together that the branches were all mixed and confused. This may give an idea of the odd and strange appearance of the root-system in a mangrove forest. Upon these arching roots, even on those which are not yet attached, multitudes of oysters may be seen. There is also a little fish (a sort of perch) which climbs up on to the roots or out of the mud below, and gasps or squatters about in it.

As to the mud itself, it is a horrible, greasy, oozy, black or blue-black slime of bottomless depth. "It is full of organic, putrefying, strongly-smelling material, clearly full of bacteria. The water itself is sometimes covered by a dirty, oily scum, and air-bubbles rising from the bottom, spread out on the surface and let loose their microbes in the atmosphere." There are many crocodiles, which may be seen reposing on the mud above high tide. It is difficult to distinguish them from a rough log of wood, but it is still more difficult to kill them, for their scales turn any ordinary bullet. There is scarcely any experience more exasperating than when, after one has taken a long, careful, and accurate aim, one observes the sleeping brute suddenly wake and scurry down into the water with a hideous leer on its face. Sea-cows or manatees are said to live in these creeks. Little ducks of many kinds rise in hundreds and thousands, but the commonest bird is the "curlew" (either a whimbrel or closely allied to it). During the day they sift the mud with their long curved beaks for insects, and at sunset fly down in vast numbers to the mudbanks near the sea. A miserable little white crane called "Poor Joe" is common, and has the same habit. It is not worth shooting, and it is quite aware of the fact. Herons, cormorants, and other birds are often to be seen. Monkeys sometimes visit the mangroves, probably to eat oysters or crabs. There are several kinds of crablike creatures which climb up the roots and may be seen running about all over them. But during the three weeks spent by the writer in the Mahéla creeks of Sierra Leone, it was the insects that made the deepest impression upon him; as soon as the evening falls the mosquitoes appear in myriads and in millions. Such creeks and mangrove swamps are always feverstricken and dangerous, and probably enjoy the very worst climate in the whole world. Of course nowadays, when Sir Patrick Manson and Dr. Ross have discovered that the mosquito carries the malaria germ, it is possible with great care to guard against malaria. One has also the satisfaction of knowing that the mosquito itself cannot be perfectly at ease with all these tiny parasites attacking its digestive organs.

At first sight such swamps appear to be useless, impossible, and dangerous. But that is not the case. No one, of course, would ever willingly reside in mangrove swamps, and the mangroves themselves are of scarcely any use to man, although the bark does sometimes furnish a useful tanning material; but, indirectly, the mangroves are one of the most important of all Nature's geographical agents.

On those horrible, slimy, shifting mudbanks no other plants could manage to exist. If one looks carefully at the seaward side of the last of the mangrove swamps, then it is easy to see that they are colonizing and reclaiming the mud.

Not only do the roots depending from the branches grasp and colonize new mud, but the seedlings are also specially adapted to fulfil the same office.

They remain a long time attached to the parent fruit; they also grow to a considerable length before they fall off. When ready to fall, they have a distinct seedling stem, which swells out towards the base and ends in a pointed root. The seedling is, in fact, like a club hanging upside down and with a pointed end. When it does fall, it goes straight down deep into the mud; then it promptly forms some anchoring roots, and the young mangrove is fixed in new mud and begins to develop. So that the forest continually grows towards the sea.

Such mudbanks soon become pierced by roots in every direction. Then the leaves of the mangroves themselves, as well as silt, soil, and rubbish floating in the water, gradually accumulate about and around these roots. This must raise the level of the ground. Eventually the soil becomes hardened and is above the level of the water. When this happens, the mangrove, which likes salt water about its roots, becomes unhealthy and the ordinary jungle trees kill it and take its place. Thus in course of time, when the jungle is cleared, fertile ricefields may be thriving on what was once a pure, or rather impure, mudbank.

In this way, by the continual development of the mangroves, enormous stretches of land are being added to the continents, and the process continues so long as the character of the coast-line favours it.

The shore-line covered by these mangrove swamps is enormous. In fact, within the tropics one finds them almost everywhere along the seashore, but coral, rock, or an exceedingly dry climate such as that of Arabia or Northern Peru, prevents their growth. Central and South America, West and East Africa, India, Polynesia, Australia, and much of the Asiatic coast-line, is covered by mangroves.

The proprietor and two of his coolies are in front of the paddy field. The young rice plants can be seen growing in the water.

Theophrastus speaks of those in the Persian gulf, and that exceedingly shrewd botanist has some valuable notes about them worth reading even to-day.

In temperate countries, such as our own, the districts where great rivers enter the sea are for the most part aguish and rheumatic, but, of course, there is nothing so startling and extraordinary as the mangrove swamps.

Yet, even in temperate countries, the work of winning or gaining new land plods steadily onwards, and it is performed by humble, inconspicuous little plants.

Where the Rhone enters the Mediterranean, there are some 40,000 acres of sandy and clayey land called the Camargue. The bare sand near the sea is often flooded and swept by violent storms in winter; anything which tries to grow there is usually carried off and destroyed.

note 75-b: Drude, *l.c.*; Schimper, *l.c.*; Warming, *l.c.*; *Colonial Reports,* No. 3, Miscellaneous. Schimper, *Indo-Malayische Strandflora.*

But, after a time, one finds here and there a solitary plant of a kind of Saltwort (*Salicornia macrostachya*) which has withstood the strain: its branches gather a little sand and hold it together, and its roots gradually explore and tie down the soil around it. Next winter it can stand the sweep and scour of the stormy water; next summer other plants begin to grow on this tiny sand-heap, and the "touradon," as it is called, is now fairly well established. It goes on growing until it may be, after a few years, six feet in diameter.

Eventually the salt gets washed out of the soil and these little heaps become united by a continuous covering of green plants in which shrubs and then trees begin to grow. By this time of course the sand has accumulated farther out to sea and the same process is going on there.

In Britain we have the "sea meadows" of Sea-grass, which covers the submerged sand and mudbanks near the mouths of great rivers.

The waving green grasslike leaves form a rich submarine meadow. They are used for stuffing pillows and cushions, especially in Venice, but their real importance in the world depends upon their being able to tie down and fix permanently those unseen shifting banks which form a real danger to all navigation.

These plants are very remarkable. They lived, no doubt, at one time on the land, like most of the flowering plants. But, like the whale and the seal, they have been driven to take refuge below the ocean. They are not easily seen, and, indeed, one may wander for years along the sea-coast and never suspect that great meadows of Zostera (the Eelwrack, Grasswrack, or Sea-grass) are flourishing under water.

But, one might ask, how is the pollen of its flowers carried? Obviously neither insects nor the wind can be of any service. The pollen of Zostera is, however, of the same weight exactly as the water, so that it neither rises to the surface nor sinks to the bottom, but floats to and fro until it reaches the outspread styles of another plant. This is perhaps the most remarkable arrangement known for the carrying of pollen.

Sometimes along the seashore, or especially on the muddy foreshore of an estuary or tidal river, one can watch those plants which are trying to form new land. One finds generally that there is a broad stretch of marshy meadow interrupted and intersected by small ditches and little winding streams. As one gets towards the shore, Sea-pink, Scurvy-grass, an Aster, and other plants, not to be found elsewhere, become common. Then stretching out into the mud there are rows of curious reeds and sedges.

Try to pull up one of these reeds, and you will find a strong, buried, stringy stem, with hundreds of anchoring roots. These are the pioneers which first fix the sand.

Over the surface of the sand between these upright stems, one often comes upon a most beautiful, glossy, dark-green, velvety cushion. It is composed of a seaweed called Vaucheria, whose twined and interlaced threads form a thick, silky cushion. But it is only beautiful to look at from above. If you pull up a piece of this cushion, you will find that it is growing on black and loathly mud, with many wriggling worms and horrible animalcula. First these pioneer reeds, then this soft, silky carpet of vaucheria, and then the sea-pinks and other estuarine marsh flowers gradually creep forward and extend over the bare muddy sand, so winning it from the sea for the use of cattle.

In the worst winter storms, when the waves are thundering heavily over these sands, it seems as if nothing could resist them. Yet if you go down when the storm is over, no harm has been done: there is the silky green cushion of vaucheria, and there are the lines of pioneer sedges and reeds quite undisturbed!

The reeds bend and sway, yielding to the water; the seaweed is slimy and oily, and the water cannot injure it. But yet the strength of these seaweeds is extraordinary, and, indeed, almost incredible.

More remarkable still, perhaps, are those seaweeds which grow upon rocks, often where the full strength of the waves beats upon them. After a heavy storm, when, perhaps, the great timbers of groins and the heavy concrete blocks of an esplanade have been shattered to pieces and tossed all over the shore, one may go down to the shore and there will be no visible difference in the kelps and tangles of the rocks. Scarcely any seem to have been broken away. Indeed, if one looks in the rubbish left by the last high tide, one finds that when one of these Alarias has been broken away, it is often because the stone itself has been torn out of the rock! One finds broken off stones with the seaweed still attached to them.

The reason is that the outside of the seaweed is oily, slimy, or slippery, so that the water gets no hold of it. The stem and substance is also elastic and surprisingly strong, so that the daily tossing and wrenching when the tides come in and go out has no effect in tearing it away.

But if you go down to a dry dock and look at the hull of a ship which has come in to be cleaned and scraped, you will see that it is entirely covered by seaweeds and shells.

That ship has been driven through the water perhaps at ten miles an hour or more, and yet those delicate-looking seaweeds have held on! It is more surprising still if you can get some of them and examine them with a microscope, for amongst them are tiny, delicate, graceful little fronds and sprays which one would think consisted of nothing but jelly. Yet they have been able to thrive and grow on the ship's hull while it has been hurrying day and night through the sea, in calm or in tempest, and in currents of hot or cold water.

Those seaweeds were called by Horace *Algæ inutiles*, or useless seaweeds; but are they useless?

Go down to a little pool and watch them waving in the water. Could anything be more beautiful than these little graceful red, yellow, or brown sprays? All sorts of seaslugs, shrimps, and minute animals of weird and wonderful design are clearly living on them. Fishes live upon these animals, and fishes are an extremely useful and excellent food for man.

CHAPTER XIII. ROCKS, STONES, AND SCENERY.

AN OLD WALL—BEAUTIFUL COLOURS—INSECTS—NATURE'S CHIEF AIM—HARD TIMES OF LICHENS—AGE OF LICHENS—CRUSTS—MOSSES—LAVA FLOWS OF GREAT ERUPTIONS—COLONIZING PLANTS—KRAKATOA—VESUVIUS—GREENLAND VOLCANOES—SUMATRA—SHALE-HEAPS—FOREIGNERS ON RAILWAY LINES—PLANTS KEEP TO THEIR OWN GROUNDS—PRECIPICES AND ROCKS—PLANTS WHICH CHANGE THE SCENERY—CAÑONS IN AMERICA.

AT first sight, and when one is striding along at some four miles an hour, there seems to be nothing at all interesting in an old wall. But if one stops and carefully examines the stones, there is a great deal that is interesting.

Rocks and walls possess a fascination of their own. Probably at least 2000 British plants are *only* found upon them, and yet of these, the vast majority are so small and inconspicuous that an ordinary person never perceives a single one of them.

It is perhaps on rocks or old walls near the sea that this stone flora is most richly developed. The nearly circular orange-yellow patches of the Lichen *Physcia parietina* are quite distinct and conspicuous. But any old wall, provided it is well out in the country, is pretty sure to be interesting.

At first it seems to have only a dull grey or neutral tint. But if one goes to four or five feet distance, one discovers that many shades of brown, red, white, and black go to make this grey.

But the extraordinary beauty of such a wall is only visible when one peers and scrutinizes the surface very slowly and carefully with the eyes six or seven inches away from it.

In doing this, one is often troubled by rude and ribald boys. A botanical friend indeed complained that he had been for months avoided and shunned as a dangerous wandering lunatic on account of his botanical enthusiasm. But true botanists get accustomed to disagreeable incidents like that, and pay no attention to the vulgar crowd.

The change in an old wall when one looks at it from a few inches distance is most remarkable. The entire surface is spotted or dusted, sprinkled or entirely covered by thick lichen stains and crusts.

The original colour of the stone is nowhere visible. The lichens show the most delicate shades and contrasts in colour; all pleasing and all blending together in harmonious general tones. The fruit of these lichens is like a minute saucer or platter generally with a thin rim or border, but it is exceedingly small, probably only one-sixteenth of an inch in diameter, or even less. The smallest of these crust lichens form continuous, very thin, coatings, covering the stone; and against this background the little saucer-like fruits show up quite distinctly.

The coating itself varies from "bright yellow, pale ochre, citron, chestnut colour, to mouse colour, different shades of grey and green, cream colour, lead colour, blue-black or pure black, tawny, brown, rusty red or pure white." The cups of one kind (*Lecidea*) are black, whilst those of *Lecanora* are generally reddish-brown. But they may be a ghostly pale hue which stands out plainly against the grey-green background of the frond.

Sometimes they are of the richest deep crimson or lake, set against a pure snow-white crust. Those of *Lecanora vitellina* are, though tiny, a brilliant yellow, and quite startling when first one notices them. Many of these contrasts and shades are never used by artists, and even from the mere artistic point of view they have great interest.

But if, after spending a few minutes in carefully looking over the rocks at a distance of six or seven inches, one stands up and goes back to four or five feet away, the whole of this colour scheme fades away and there is only the monotonous indeterminate grey or neutral tint of the wall.

Now why is this? Why should these delicate and exquisite shades be wasted on such minute and scarcely distinguishable forms?

There are always two sides from which one can look at any subject, namely the *inside* and the *outside*.

From the inside (that is from the point of view of the little lichen itself) these colours are decidedly useful. Small insects crawl about on such walls or hover a few inches in front of them, and to those insects these cups will be as conspicuous and attractive as a scarlet geranium is to ourselves.

Just as we habitually go to look at a geranium, so those insects fly towards the cups and crawl about on them. Then when the spores and dust of the lichen begin to stick in their hairs and feet, they go to a bare place and clean or brush them off. Thus the spores and dust are carried to a new part of the rock, where they will grow if they can find an unoccupied place. The taste in colour of these insects, moreover, is apparently not very different from that of man.

But perhaps a still more interesting point of view is that from the outside. Why are those lichens there? What are they doing, and are they of any use?

The general scheme of Nature is to cover the whole world with green, so that every ray of sunlight may find a working leaf or green frond ready to welcome it and use it. Nature abhors bare rock, barren sand, and empty water, and never ceases to try to bring it under that beautiful covering of green plants and active vegetable life which supports both man and animals.

We all know that there is a romance in the story of man's colonies. First the explorer searches out the country; then the pioneer frontiersman settles and builds his log-hut or rough shanty. Next comes the frontier village, which may perhaps in many years' time become a crowded city where active, valuable work is carried on.

The story of the colonizing of rocks and stones by plants is just as vividly interesting. These tiny lichens are almost the first pioneers, and prepare the ground for those that follow. Upon that bare rock, life is terribly severe. The frost shatters it, sunshine heats it until it almost burns the hand in summer. Floods of rain or of sleet beat against it, and it may be frozen over for weeks.

What plant can stand such conditions? Only these minute, tiny, scarce visible lichen films!

Gradually new lichen crusts develop upon it. They cover over the first pioneers; first they suffocate them and afterwards devour their remains. Nature is very businesslike and severe in her working. The lichen crust may be now about one-sixteenth of an inch thick. It is a very slow process. There is a story of a boy who noticed a patch of lichen near his father's door. He went away to Kamschatka or somewhere and came back a very old man of eighty-five years; but he found that the lichen patch was just the same size as when he went away. That, however, is just a story!

At any rate, one of these little crust-lichens called *Variolaria* has been known to increase half a millimetre in size (about a sixtieth part of an inch) between the end of February and that of September.

Now if one tries to realize what the life of such a lichen crust or crottle must be, it is obvious that the stone below it must be a little corroded or weathered, and remains of the first choked pioneers, bacteria, and possibly tiny insects or animalcula will be under the crust, which may now be one-sixteenth of an inch thick.

It is the turn now of other lichens to colonize it. These may be the little trumpet or horn and cup lichens, *Cladonias*, or perhaps the larger grey kinds, *Parmelias* and *Physcias*, which have leaf-like fronds and form circles of perhaps eight to ten inches in diameter. The crust-lichen is overgrown, broken up, disorganized, and devoured by the *Parmelias* and *Cladonias*, who are helped by bacteria, insects, and animalcula which shelter below them. These leafy lichens grow much more rapidly.

They may increase two-thirds of an inch in one year.

But very soon after this, one notices a few inconspicuous green mosses; at first in crevices between the stones or in hollows, and not remarkable, they soon increase and form trailing sprays or branches which grow very quickly. Branches of moss four or five inches long extend over the leafy lichens in a season. The *Parmelias* and *Cladonias* struggle on, but they cannot keep pace with the rapid life of the moss, and soon our wall is covered by beautiful moss turfs.

Underneath such a turf there may be an inch or so of good soil (dead moss and dust with lichen and insect bodies). Worms, insects, etc., shelter and flourish and multiply in this soil.

But the turn of the moss is coming. Here a few grass-blades, there a tiny plant of Sandwort, possibly a Rock Bedstraw, begin to root themselves in the moss.

If people would only let the wall alone, it would soon be festooned with hanging plants, and producing quantities of grass, but somebody is sure to find that it looks very untidy, and everything is torn off the wall, which again looks new and raw and clean. Then of course the pioneer lichens begin again!

Some very interesting and remarkable facts have been discovered about the way in which lavas and basalts have been occupied by the plant world.

In the great volcanic eruption of 1883, the whole island of Krakatoa was covered by hot lava and glowing ashes. In 1884 and 1885 the sunsets were remarkable for a curious fiery red or orange glow, which was popularly supposed to be due to the volcanic dust of that explosion. It is said that the dust travelled three times round the earth, though I do not know on what authority.

However, on Krakatoa island there was left a clean "slate." There were neither bacteria, nor leaf-mould, nor living plants of any kind; no spores or seeds could have endured the fiery furnace of the eruption.

Three years afterwards the botanist Treub visited the island. He found that the rocks had been first covered by thin layers of minute freshwater Algæ, but that ferns were then occupying and inhabiting the lavas. Eleven kinds of ferns, and but very few other plants, were discovered.

People were interested in this, and Dr. A. F. W. Schimper then visited another volcano which had been pouring out huge streams of lava in 1843. He found that there were still plenty of ferns, but also numbers of shrubs and other plants. Yet even then there were no trees, and there was no continuous mantle of green plants such as we are accustomed to in this country. He also found many plants growing on the lava which are generally found on the branches of trees, that is, which can do without a thick layer of soil. He also found quantities of a pitcher plant, Nepenthes (which lives mainly on insects caught in its pitchers).

This does not at first sight seem to agree at all with what has been given for the walls. It is true that sometimes in the Highlands, or Lowland and Lakeland Hills, one comes across quantities of the Bladderfern and others growing on the "screes." (These last may be described as streams of broken, angular stones, filling small gullies, and spreading out at the base over a considerable space.) Often these ferns seem to be all that can thrive in amongst the stones. But in a mild and temperate country like our own, one would expect things to proceed differently.

And in fact they do so. Every one must have noticed a green stain which covers wet walls, stones, stucco, even marble statues, and especially tree bark in wet or damp situations. This is a minute green seaweed rejoicing in the name of Pleurococcus. It is a pretty object for the microscope.

This, of course, is the first stage of colonization. It is followed by mosses of sorts.

But there is a more interesting series still in a climate resembling our own. The lava-flows from Mount Vesuvius have been investigated by several observers.

There it was found that the first inhabitants were *lichens* and small green seaweeds; then "different mosses occupied the lava over which a certain quantity of vegetable dust had been scattered." After this, scattered ferns and even small shrubs could be seen even on flows which were red-hot only twenty years before, whilst on old lava-fields herbs, shrubs, bushes, trees, and even true woods had developed.

Yet in Greenland lava-flows dating from 1724-29 are still only covered by crust-lichens and a very few of the stone-mosses! In Sumatra, on the other hand, the volcano of Tamboro, which in 1815 had entirely destroyed its vegetation, was covered with a fine young wood in 1874! The strong heat and abundant moisture of Sumatra favours, whilst the horrible climate of

Greenland prevents, the rapid growth of good soil. Just as cities of 20,000 inhabitants can spring up in a few months in the Western United States, whilst the Esquimaux of Greenland have not managed as yet even to live in villages!

The full beauty of this gradual colonization and occupation of bare rock and stones only impresses one properly if one tries to trace the stages, but it is an interesting history.

Near Glasgow one sees great heaps of shale or blaes (generally *blackband*), which are often mistaken for natural hills. This is or was virgin soil, never occupied by plants, and entirely destitute of leaf-mould or any sort of organic plant-food.

If one scrambles to the top of one of these heaps, it is easy to see all the details of the occupation. Long underground runners of coltsfoot and of horsetail are climbing up the sides, fringes of creeping buttercup, couchgrass, and other hardy weeds occupy, every year, a little more of the flanks, but, on the top, one very soon finds that the dust of the atmosphere, aided by weathering, has afforded a chance to mosses, to hawkweeds, and other rock plants. These in time cover the top, and soon hardy grasses and weeds form a regular turf on the top of the shale.

It is interesting to scramble to the top of one of these heaps, especially in summer. One then begins to realize how every plant attends strictly to its own business.

All over the sides of the heap there will be hundreds of a rare groundsel (*Senecio viscosus*), which is not really a native, and *never* occurs except on such places. In a grass field close by hundreds of thousands of Ragwort (*Senecio jacobæa*) make a glorious golden carpet; in the marshy part of the meadow the Water Ragwort (*Senecio aquaticus*) may be found. In the cottage gardens and here and there along the roadside the groundsel (*Senecio vulgaris*) is flourishing abundantly.

These plants never interfere with or encroach upon one another's grounds. Every year thousands of ragweed and groundsel seeds must be blown on to the shale-heap, but they never manage to grow there.

It is only the foreigner (*S. viscosus*), accustomed to a very hot and dry climate, and with sticky leaves which catch atmospheric dust and probably insects, that can exist on the bare shaly sides. These slopes of shale are easily heated by the sun, and at the same time radiate the heat rapidly away, so that the Viscid Groundsel must have a very hard time of it. When its roots have worked up the shale a little, and its dead leaves have covered the surface with mould and organic matter, then possibly others (true British plants) can get a footing and suppress it.

Along railway tracks, also, the ballast forms a very hot, a very dry, and a very barren soil. Many of the regular railway-track plants are foreigners from the far south, even from the sunny shores of the Mediterranean. They are mostly annuals, such as the little Toadflax (*Linaria minor*), which can just manage to exist under those conditions.

Of course, the sides of the banks and of cuttings on railways are generally formed of good earth or soil, and support a rich and flourishing flora of true Britons.

Besides these slow, laborious lichens, mosses, and others which attack rock, there are other plants which are generally called rock plants, though they behave quite differently.

These are those fine hardy Hawkweeds, Roseroots, Sempervivums, Mew, and others which establish their roots in cracks or crevices of the rocks.

Such cracks are soon full of good soil, for the wind blows decayed leaves and dust into them, and the roots are always burrowing into, eating into, and shattering the rocks. Most of them have a circle of leaves which are pressed flat to the ground. Thus they escape the violent winds and storms always common on such crags and precipices. The flowers, however, supported on tough, strong, and flexible stalks, sway freely to and fro in the wind, and can be seen by insects a long way off.

These rock plants are of some importance as stonebreakers and pioneers in a very interesting process.

Wherever a cliff or precipice of stone is exposed, it is "weathered." Water gets into the cracks and freezes in winter. But when water is frozen it expands or widens, and as this happens to the water in the crevices and cracks of rocks, pieces of rock are shivered and broken off. Besides frost and wind and rain, these rock plants help to attack the cliff. Their roots get into the crevices, and there widen and expand, tearing off great slabs and splinters of rock which fall down to the foot of the cliff.

Down below plants are every year growing over and covering up or "happing up" with green these bare fragments and splinters. A considerable amount falls down every year, so that the ground is always being raised up below the precipice. At the brow or edge above the precipice, there is also always a loss of rock and stone every year.

So that every year the bare rock exposed becomes smaller and smaller, until eventually a steep, green, grass-covered slope covers over the entire site of that precipice.

Moreover that is not by any means all that plants do in the way of changing the scenery of the country. Look at the outlines of the hills in any part of Great Britain except in the broken, jagged, rocky mountain ranges of Scotland and Wales (also Cumberland, Westmorland, parts of Derbyshire and Dartmoor tors). Everywhere there are smooth, flowing, gently undulating rises and falls. No sharp, abrupt descents break these graceful sweeping curves. If you compare the scenery of a cañon in the rainless deserts of Western America, the contrast is very striking. There the sides of the valleys are steep cliffs; it is all harsh, precipitous, horrible country, which is obviously very unpleasant and very unattractive to civilized people.

It is this green covering of plants which makes the difference. The rain that falls is not allowed to cut out ragged ravines; it is intercepted and soaks into the grasses, which so keep a smooth, gentle outline over hill and valley.

If you notice the effect of a heavy shower of rain on a road or bare earth, you will see how soon tiny valleys and cañons and beds of streamlets are cut out. But on the green fields beside the road, there is no change in the surface at all! It seems to be quite unaffected by the heaviest storm of rain.

CHAPTER XIV. ON VEGETABLE DEMONS.

ANIMALS AND GRASS—TRAVELLERS IN THE ELEPHANT GRASS—ENEMIES IN BRITAIN—CACTUS *VERSUS* RATS AND WILD ASSES—ANGORA KIDS *V.* ACACIA—THE WAIT-A-BIT THORN—PALM ROOTS AND SNAILS—WILD YAM *V.* PIG—LARCH *V.* GOAT—PORTUGUESE AND ENGLISH GORSE—HAWTHORN *V.* RABBITS—BRIERS, BRAMBLES, AND BARBERRY—THE BRAMBLE LOOP AND SICK CHILDREN OR AILING COWS—BRIERS OF THE WILDERNESS—THEOPHRASTUS AND PHRYGIAN GOATS—CARLINE NEAR THE PYRAMIDS—CALTHROPS—TRAGACANTH—HOLLIES AND THEIR INGENIOUS CONTRIVANCES—HOW THORNS AND SPINES ARE FORMED—TASTES OF ANIMALS.

BY far the greater number of wild animals live by eating vegetables. If one thinks of the elephant's trunk, the teeth of a hippopotamus, or even of the jaws and lips of mice, rats, and voles, the thoroughly practical character and efficiency of their weapons become the more astonishing the more one reflects upon them.

Yet the defences adopted by plants are just as wonderful, and are often most ingenious.

It seems at first remarkable that the most usual food of animals, grass, should be apparently unprotected. It is upon grass that the great herds of bison, of buffalo, of antelope, and guanaco, are or were supported. Yet grass is so wonderfully reproductive, produces such enormous quantities of buds and foliage, and grows in such luxuriance, that there is no fear of its being killed out.

There are many places in the world where vegetation defies the attacks of the animal world. Neither man nor elephant can live comfortably in the thick jungles of West Africa and the great forests of Brazil. Nor can either man or elephant utilize great tracts of country in Central Africa which are covered by the Elephant Grass.

For, perhaps, four or five hours the weary caravan plods on through a sort of burrow, two feet wide, made in this gigantic grass. The stems are ten feet or more in height, and nearly meet overhead. There is nothing whatever to be seen except the narrow path. The atmosphere is stifling and hot. To cut a new road a few hundred yards long through it involves hours of labour. It is only when there has been a long drought that it is possible to set fire to the Elephant Grass, and then for a very short time the young growing shoots can be grazed. But no cattle can break through when it is fully grown.

The very exuberance of vegetation in such cases prevents any harm.

Perhaps it is best to show how, even in Great Britain, all plants have many dangerous foes. The roots of trees are nibbled by mice, voles, and sometimes by swine. The bark is injured by cattle, roedeer, reddeer, fallowdeer, who tear the bark with their horns, and especially by rabbits and hares. The leaves are eaten by the same animals and also by horses, goats, sheep, etc. The young buds are attacked by squirrels, who also break off the leading shoots of certain firs when they happen to be in a playful mood.

But it is in cultivated lands and in open, rather dry and arid country that one finds the most interesting weapons in the fights between plant and animal. It is in such places that some of the most beautiful and useful creatures have their home. The horse, ass, camel, goat, and sheep probably belong to those wonderful lands which border the great deserts of Africa and Asia. These animals have been obliged to travel far and fast, and to perfect their bodily strength in order to pick up a living.

These giant grasses are sometimes one hundred and twenty feet high and one foot in diameter. They at times grow at the rate of three feet per day, and are used for all sorts of purposes, such as scaffolding poles, flower-pots, as a vegetable, etc. etc.

They have been taught (perhaps we should say learnt) by the thorns and briers of the wilderness.

The Cactus, Prickly Pears, or other succulent plants which belong to true deserts, are covered over with most curious and interesting spines. A row of little projections runs down each edge of the round fleshy stem. On each projection there is a rosette of spines. Sometimes these are long, slender, and diverging; in other cases they are short, stout, and curving over.

Now imagine a guanaco in South America, or even a rat or mouse, which is perishing of thirst in the arid desert where such things are found. It will be seen that it is by no means easy for it to taste the water in the juicy stem, for even the thin muzzle of a rat could scarcely get between the thorns.

Kerner describes how the wild asses in South America root up or try to split the Cacti with their hoofs to get at the juicy tissue of the unarmed lower parts. Yet they often receive dangerous wounds in doing so from the frightful spines of Melocactus and others.

It is very interesting to see a flock of Angora goats in South Africa attacking an Acacia. The kid is a pretty, white, fluffy little creature, with the most meek, mild, and innocent expression. Yet it is a quarrelsome little brute. In a few minutes an Acacia will be despoiled, broken, and robbed of its foliage by a flock of them, although it bristles all over with long spines, of which there are a pair at the base of each leaf.

Even the *Kameeldorn*, Camelthorn Acacia, or the Wait-a-bit in South Africa cannot defend itself.

The Wait-a-bit (*Wacht een beetje*) is so called from the ingenious nature of its spines. There are two together, of which one is straight and the other curved round like a hook. Both are very sharp and strong, so that an incautious traveller is sure to injure himself and his clothes. The straight one runs into his tender flesh, whilst the curved one fixes itself in his clothes.

It is by thorns, spines, and prickles that plants often protect themselves against the attacks of grazing animals. But it must be remembered that these are by no means the only safeguard. Plants produce poisonous, bitter, or strong-smelling substances which keep off their enemies, and these indeed often afford a more efficient protection (see Chap. III.). These thorns, etc., can be produced in the most unexpected places. There is one rule, however, namely that they are invariably found in the exact spot where they can be most useful.

Thus there are certain palms which possess green, juicy leaves, much relished by snails. These are protected by a sort of spine entanglement formed upon certain roots, which grow at the base of the leaves. Nor is this the only case in which spines are found on roots. There are certain South African bulbs (*Moræa*) which are protected from the wild pigs by a dense mass of spiny roots. On my march to Uganda from Mombasa, I was very much astonished to see an extraordinary Wild Yam. It had a huge underground tuberous part full of starchy matter, but it was quite impossible for any marauding wild boar to get at it, for it was entirely enclosed in a sort of arbour of long, arching roots densely covered by stout spines, which made a perfect protection.

It is more usual to find thorns developed on the branches or stems. Generally these are formed on the outside towards the end of the branches. In the Alps, larches have to suffer from the attacks of goats which nibble off the ends of the young shoots. The part behind the scar dries up, but fresh twigs are put out from further back along the branch, until the tree becomes a closely branched, twiggy, bristling mass which looks like the clipped yews in old gardens. But so soon as it has grown tall enough to be above the reach of the goats, an ordinary larch stem develops and may grow into quite a respectable tree. This fact is given by Kerner von Marilaun (*l.c.*, p. 445), and is very instructive, as explaining why it is that so often the ends of the branches become hard thorns: the green leaves and twigs are hidden and protected.

One of the neatest examples of this is the Portuguese Gorse or Whin, which resembles a little cushion with every branch ending in strong thorns and every leaf terminated by a stout spine.

The common Whin, Furze, or Gorse, is very nearly as perfect an example of thorniness and spininess. The Southdown sheep do not seem to injure it on those beautiful Sussex downs so famous for succulent mutton, yet in the early spring, or in a very wet season, one often finds in the grass at the foot of the bush (or even in the bush itself) small shoots which would be taken at first sight as belonging to some other plant. These little shoots are grey with hairs and have soft trefoil leaves which are quite unprotected, for their spines are quite soft. They are probably seldom eaten, for most of them are in the shelter of the old spiny bushes.

Yet even the old bushes can be used as fodder for sheep if they are crushed and ground up so as to break the thorns and spines. The Gorse is a very hardy plant, and is said to be only out of flower "when kissing is out of fashion".

There is still some uncertainty as to the exact way in which animals set to work when they are eating thorny or spiny bushes. This makes the arrangement of the thorns sometimes a little difficult to follow. Moreover it is often not so much the leaves as the juicy bark in winter and early spring that is required. Sometimes everything above ground is eaten down.

Rabbits, for instance, do not as a rule touch the Hawthorn, yet Mr. Hamilton says, "The second winter after planting was very severe and this hedge was eaten down to the very ground by rabbits. For about 600 yards I do not think that a single plant was missed." In frost and snow almost every plant is attacked by rabbits, and indeed by any grazing animal.

Remembering that it is very often the young juicy shoots that are sought after, it is quite easy to see why the young Rose suckers and shoots from the base of the stem fairly bristle with long and short prickles. These latter are generally straight, not curved like those of the long arching branches which are supposed to hook themselves on the branches of the surrounding trees. The young light-coloured branches of the cultivated Gooseberry are flexible, and hang over in such a way as to make it difficult for an animal to reach the bark: a cow or sheep, if it wished to eat these branches, would begin at the hanging tip and make a sort of upward tearing jerk while its tongue gathered the branch into its mouth. If one copies this with the hand it is easy to see how the length and arrangement of the prickles and the flexible nature of the spray would make such a proceeding on the cow's part most uncomfortable.

So also in the Barberry, the young juicy upright shoots which spring from the older branches have stout three to seven-branched prongs pointing downwards, of the most efficient character. Each is really a modified leaf and is found below each bud. Even the mere idea of an animal's tender lips or tongue tearing at these shoots from below gives one a momentary shudder. In the younger, wavy branches of the Barberry the spines are straighter or more diverging. The young leaves of the short bud above alluded to are also most efficiently protected by their spines. The Hawthorn has a curious arrangement of very long stout thorns, behind which the leaves are sheltered. The younger flexible branches have smaller spines, which become efficient in winter and tend to prevent animals from eating the bark. The Cockspur thorns are 4 to 5-1/2 inches long, and extremely like the spur of a gamecock.

Bramble prickles are generally curved back in order to hook or cling to the branches of other trees, but any one who has tried to force his way through a clump of brambles knows the difficulty of doing so. The loops made by the branches fixing themselves in the ground were at one time given credit for healing various diseases. Children in Gloucestershire used to be dragged backwards and forwards under these loops; in Cornwall also people afflicted with boils were made to crawl under them. Even cows when suffering from paralysis (supposed to be due to a shrew-mouse walking over them) were dragged through the Bramble-loop, in which case Professor Buckman remarks, "If the creature could wait the time of finding a loop large enough and suffer the dragging process at the end, we should say the case would not be so hopeless as that of our friend's fat pig, who, when she was ailing, had a mind to kill her to make sure on her." The brambles and briers of Gilead and Ezekiel were probably brambles of which *Rubus discolor* is common in Palestine, and the Butcher's Broom (*Ruscus aculeatus*). This last plant is really of the Lily family, and its flat leaf-like branches end in a sharp spine. The rabbit does not eat it.

Amongst foreign thorny and spiny plants it is very difficult to make a selection.

Theophrastus (one of the very earliest botanists—see) describes a class of shrub very common in Phrygia, in which the leaves are produced at the base of the young shoots, which latter end at the top in branch thorns. These thorns, therefore, entirely cover the foliage and keep off that vegetable demon the goat. Some of the Crucifers, Roses, Composites, Labiates, etc., take on this habit in goat-infested countries.

In Egypt, near the Pyramids, one often finds *Carlina acaulis*, a little thistle which has no stem, but is merely a flower seated in the middle of a rosette of leaves which lie flat on the sand. In the centre there is a circle of sharp spines, each of which is from one to two inches in length. The nostril of a hungry camel or donkey is sure to be pierced if it tries to eat the leaves. The spines of this thistle, like those of our Carline and the *Centaurea calcitrapa* (thistle of the Bible), spring from the bracts surrounding the flower.

The ancient "calthrops" or "crawtaes" (first used by the Romans) were designed from the spines of the last-named plant (*calx*, heel, and *trappa*, snare.) It had four iron spines, so that, however it was thrown down on the ground or in a ford, a spine was sure to stick up and to lame man or horse

1. Old Roman Calthrops, left on roads, fords, etc., to lame horses. 2, 3. Fruits of Tribulus, showing efficient spines. Animals' feet, in passing, must catch them. They are more efficient than Calthrops.

The Tragacanth plant has also very neat spines. They are the persistent spiny stalks or midribs of the older leaves from which the leaflets have dropped away. The fresh green leaflets are quite protected inside these withered spines.

Several grasses have leaves which end in sharp or needlelike points. One of these, *Festuca alpestris*, actually produces bleeding at the nostrils of grazing cattle, and is detested by all the shepherds of the Alps.

The Holly is one of our most beautiful trees, as John Evelyn points out: "This *vulgar* but *incomparable* tree.... Is there under *Heaven* a more glorious and refreshing object of the kind than an impregnable *hedge* of near *three hundred* feet in *length*, *nine* foot *high* and *five* in *diameter*: which I can show in my poor *Gardens* at any time in the year, glittering with its arm'd and vernished *leaves*? The taller *Standards* at orderly distances blushing with their natural *Coral*." This apparently was the identical hedge into which Peter the Great used to trundle his wheelbarrows. The barrows contained his courtiers. There was a nice run from the top of rising ground close at hand. It was at Sales Court, Deptford.

The spiny leaves of the Holly are unfortunately not nearly strong enough to save it from its enemies. The bark is apparently of a particularly delicious and toothsome nature, for sheep, cattle, and the ubiquitous rabbit are always delighted to destroy the trees.

It has been noticed that wild hollies have at the base very spiny leaves, but that higher up on the tree (above the reach of cattle) the leaves have no spines at all. Sir Herbert Maxwell, in his *Memories of the Months*, takes up this question. It is best to give the description in his own words:—

"I strolled out along the banks of Tay in that noble woodland which is continuous from Dunkeld to Murthly. Here there are many fine hollies, some on the river banks and cliffs, others on level ground, planted by no hand of man. There was not one of these which did not confirm my observations first made many years ago, and hardly one which did not bear evidence of special growth—not merely as a reaction against pruning or cropping, but *as a precaution against any such contingency*—so regular and deliberate as to suggest that these trees are something more than unconscious automata.

"Many of these hollies are thirty feet high, with foliage down to the ground. They carry spinous leaves up to a height of three or four feet; above that level all the foliage is absolutely smooth and spineless. One tree rose from the ground in two bare stems, and the lower branches did not reach below the browsing level. But from between the two old stems rose a young shoot about four feet long, clothed throughout its entire length with intensely prickly leaves. This tree was growing in an enclosed wood where cattle could not come; still, roedeer might be about, and the holly armed its young growth at the low level, although the leaders of the old stems, not less vigorous in growth, bore leaves as smooth as a camellia's. I noted one particularly suggestive tree, an unhealthy one. The growth had died back along most of the branches, which stood out bare and dry; but a recuperative effort was in progress; fresh and luxuriant growth was bursting along nearly the whole height of the stem, and the foliage of this was vigorously prickly up to about four feet, and smooth above that height. I noticed many instances of localised prickly growth where boughs, originally above the browsing level, and clothed with spineless leaves, had been weighed down and cropped by cattle. But this is merely a vigorous reaction against external injury, such as makes a clipped holly hedge bear spinous foliage from base to summit."

This quotation shows that there is no doubt as to the facts. It is true that one finds cultivated hollies showing many variations. Sometimes all the leaves are spiny, both above and below. In other varieties none of the leaves possess spines at all. Yet it must be admitted that these are facts and cannot be denied.

Moreover, the Osmanthus, with its holly-like leaves, the Evergreen Oak, and some Junipers are found to show exactly the same curious difference. The perilously-situated lower leaves are more spiny than those which are above the reach of grazing animals.

Kerner von Marilaun also has remarked a similar protective arrangement in *Gleditschia chinensis* and in the Wild Pear. Trees of the latter, when they are young, "bristle with the spines into which the ends of the woody branches are transformed"; but tall trees twelve to fifteen feet high are entirely without thorns!

It is when one meets coincidences of this nature that the full meaning of plant life begins to dawn upon the mind.

How is it that the plant knows the time to produce its spines, and the time to refrain from doing so?

There are certain queer facts that have been given on good authority as to the causes which tend to produce thorniness and spininess.

Linnæus, *Philos. Bot.*, p. 215, § 272, says:—

"Spinosae arbores cultura saepius deponunt spinas in hortis." Lothelier found that Barberries grown in a moist atmosphere had no spiny leaves, and that the thorns were far less woody under those conditions, whilst in a perfectly arid and dry atmosphere only spines were formed; a strong light also tended to produce spines.

Professor Sickenberger grew a desert plant (*Zilla myagroides*) in the Botanic Garden at Cairo, and found that its spines were much weaker and more slender than the strong rigid thorns which cover it in its natural desert.

Professor Henslow found that the spiny form of the Rest Harrow, when grown in a rich soil with an abundance of water, gradually loses its spines. All these experiments certainly show that a dry desert sort of life, and possibly strong sunlight, favour the development of spines and thorns.

Of this there cannot be any reasonable doubt, for the extraordinary quantity of thorny, spiny things in deserts shows that there must be some connexion between such a life and their production (see Chapter X.). In such places animals are always abundant. But these hollies, pears, and other plants show exactly the opposite to what we should expect. It is when the head of the young holly reaches the sunlight and feels the wind that its leaves become harmless!

If one remembers the case of the young larch and its goat enemies on page, it is perhaps possible to think that the lower branches and twigs were for untold generations exposed to laceration and biting. Thus, suffering from the loss of water by these regular annual wounds, the leaves developed their spines in response. So far, belief is not more difficult than it is with regard to the origin of any variety. But whenever, by reversion to their ancestral type, the original not-spiny leaves developed on the top of a tree, that tree would have an advantage, for every leaf on it would be more economically produced; a smooth leaf would not require to spend food in order to make spines. Such trees, spiny below and smooth above, would be best fitted to survive, healthier and more vigorous, and in the end would leave more descendants.

At the same time, such a case as this reveals again that mysterious and exquisite purposefulness which a reverent mind discovers in Nature everywhere.

At the same time, as we have already pointed out, we are exceedingly ignorant of many of the very commonest facts. Léo Errera, the great Belgian botanist (whose recent death has been a terrible loss to science), collected together some facts as to the taste of cattle for various spiny and thorny plants; he found that cattle wished to eat the following: Buckthorn, whin or gorse, raspberry, brambles, the Scotch thistle, the creeping thistle, as well as musk, welted and slender thistles, sow thistle, and saltwort.

They avoided: Barberry, the petty and German whin, rest harrow, the carline, and the other thistles not given above, as well as the common juniper.

They disdained or despised: Sea holly, common holly, milk thistle, *Lactuca*, and *Urtica urens*.

So far as the holly is concerned, it is certainly not despised by sheep and rabbits in this country. But how few are the plants investigated! Several of the commonest British plants are omitted just because no one has taken the trouble to watch them.

Here, then, is an opportunity of discovering something new, fresh, and interesting which should be well within the reach of any one who passes his life in the country.

CHAPTER XV. ON NETTLES, SENSITIVE PLANTS, ETC.

STINGING NETTLES AT HOME AND ABROAD—THE USE OF THE NETTLE—SHAM NETTLES—SENSITIVE PLANTS—MECHANISM—PLANTS ALIVE, UNDER CHLOROFORM AND ETHER—TELEGRAPH PLANT—WOODSORREL—HAVE PLANTS NERVES?—ELECTRICITY IN THE POLAR REGIONS—PLANTS UNDER ELECTRIC SHOCKS—CURRENTS OF ELECTRICITY IN PLANTS—THE SINGING OF TREES TO THE ELECTRO-MAGNETIC EAR—EXPERIMENTS—ELECTROCUTION OF VEGETABLES.

THE common nettle is one of our most interesting British plants. It is exposed to great danger; one sees it growing not only in pastures and parks, but in waste places, along roadsides, and near cultivated ground. Yet it is very seldom either eaten or even touched. Cattle do occasionally eat the young shoots. But this is exceptional, for even in fields where there are plenty of cattle great clumps of nettle luxuriate and increase in size every year.

The stinging hairs are hollow and shaped rather like a narrow bulb or flask; the tip is slightly bent over and rounded (not sharp); the hairs contain formic acid. If one grasps the nettle or strokes it in a particular way (from below upwards) the hairs are pressed flat against the stem or broken, so that no wound is made by them in the skin and consequently they do no harm. But if the point of the hair pierces the skin, the well-known irritation is set up. That is because formic acid is poured into the wound. Besides the stinging hairs which keep off all the larger animals (including man) there are others, shorter and thickly set, which do not sting at all, but are intended to keep off snails.

The pain produced by our common nettle is, however, a very trifling matter compared with that produced by some of the foreign species. One of the Indian kinds was used to excite and irritate bulls when they were intended to fight with tigers in the games which used to be held at some Indian Courts. Another found in Timor is called the Devil's Leaf; the effect of its sting may last for twelve months and may even produce death. But a still more dangerous stinging plant is a handsome tree (*Laportea moroides*) found in Australia. It is often 120-140 feet high, and has fine dark-green leaves often one foot in length. The sting is so powerful that even horses are killed by touching its leaves. The sting of *Jatropha urens* is so strong that people become unconscious. In Java also the sting of *Urtica stimulans* continues to smart for twenty-four hours, and may produce a fever which is very difficult to shake off.

Yet our common nettle is the favourite food-plant of the caterpillars of the Small Tortoiseshell, Red Admiral, Peacock, Camberwell Beauty, and other butterflies. These caterpillars are possibly more intelligent than many of our country folk, who do not know that the nettle is a very useful plant, as the following statements most clearly prove. Its young leaves make an excellent spinach, and it was, according to Sir Walter Scott, formerly cultivated in Scotland as a pot-herb. Pigs, turkeys, geese, and fowls like the leaves when they are chopped up. It is said that the dried leaves and seeds will make hens lay in winter time. The seeds, under pressure, yield quite a good oil. A yellow dye can be obtained by boiling the roots with alum. An excellent string can also be made from the inner bark of the stems, which has, in fact, been used to make twine and even clothing. The nettle is also valuable as an external stimulant in cases of paralysis.

A plant with so many wonderful properties would not be so common as it is, or so little disturbed, if it were not for its powerful stings.

There are one or two plants which are extremely like the nettle at first sight. Lord Avebury has an illustration in his excellent little book in which it is most difficult to tell which are White Deadnettles and which are stinging nettles. No doubt the harmless deadnettle is helped to escape injury by this resemblance. The Hemp Deadnettle and some Campanulas are also very like it when growing. These also are sham nettles and may escape in the same way.

There are several common greenhouse Primulas which also produce irritation of the skin. When handled by gardeners a painful smart is set up which lasts for some time. *Primula obconica* is the worst of these, but *P. sinensis*, *P. cortusoides*, and *P. Sieboldii* sometimes have the same effect. In all these cases it is due to a peculiar secretion of certain glandular hairs.

The methods of protection against grazing animals so far described, such as stinging hairs, thorns, spines, etc. are obvious enough, but perhaps the most ingenious system of defence is that exhibited by the Sensitive Plant and a few others.

When man or any heavy animal is approaching certain Indian plants, their leaves suddenly drop, and the leaflets close together. The mere shaking of the ground or of the air produces these extraordinary movements in the sensitive Woodsorrel (*Oxalis sensitiva*), in two Leguminous plants (*Smithia sensitiva* and *Aeschynomene indica*), and in several Mimosas.

When one leaf-tip of *Mimosa pudica*, the Sensitive Plant (*par excellence*), is touched or injured, a series of changes begin. All the little leaflets shut up one after the other; then the secondary stalks drop; after this the main stalk of the leaf suddenly droops downwards. After a short interval, the next leaf above goes through identically the same movements. If the shaking or injury is severe, every leaf from below upwards moves in the same way.

One probable advantage of these movements can be understood from the behaviour of flies, which alight upon the leaves and make them drop. The flies are startled and go away. Grazing animals will consider such behaviour in a vegetable as very uncanny, and will probably go to some other less ingeniously protected plant.

Of course such extraordinary behaviour has been a challenge to the botanical world, and there is an overwhelming mass of speculation, and observations about the Sensitive Plant.

It has been proved that the movements are caused by the thickened part at the base of the main stalk of the leaf. This is swollen, and full of water, and much thicker than the stalk itself. It is by this thickened portion that the leaf is kept at its proper angle. When the tip of the leaf is shaken or injured, the cells on the under side of this swollen part allow their water to exude into the spaces between them, and in consequence down comes the leaf-stalk.

This is not, by any means, a full or even a sufficient explanation. There is certainly some peculiar sending of messages from the tip of the leaf to the swollen part itself. It is not safe to say that it is a nerve message, but the process resembles the way in which messages are sent by the nerves in animals. Not only so, but the contraction of the under side and a corresponding expansion on the upper side, resembles the muscular movements of contraction and expansion in animals.

It must always be remembered that plants are alive; their living matter is not in any way (so far as we know) essentially different from that of animals or of man. Their living matter (protoplasm) in leaf-stalks and leaves is cut up into boxes or cells, each enclosed in a case or wall of its own. Yet these are not entirely independent and unconnected, for thin living threads run from cell to cell, so that there is an uninterrupted chain of protoplasm all along the leaf, leaf-stalk, and stem.

In this particular case of the Sensitive Plant, the leaves at night regularly take up the position which they adopt when injured or shaken during the daytime.

The easiest way to produce the shrinking of the leaves is, as has been mentioned, to hold a lighted match a little below the leaf-tip. Severe shaking, a strong electric shock, or a railway journey will also produce closing of the leaves.

Under chloroform or ether, or if the atmospheric pressure is suddenly diminished, the leaves will also fall. In some respects they are very lifelike, for if too often stimulated they become "fatigued," and will not react unless a sufficient interval of rest is allowed them.

The reaction occurs very soon if the plant is in good condition: in less than one second it begins, and the leaf-stalk may fall in two to five seconds, but the recovery is very slow.

Vivisection is a cruel sort of proceeding, although it may sometimes be necessary. The most curious vivisections have been performed on Mimosa. When the leaflets are cut off, it is possible, on a stimulus being applied, to see water oozing out of the cut surface of the stalk. This would go to show that it is the water being discharged from the leaf-base that produces the movement.

There are, however, many points in the behaviour of the Sensitive Plant which have not yet been explained.

Possibly the curious Semaphore or Telegraph Plants, whose leaflets suddenly and without any obvious reason move with a jerk through an angle of several degrees, may also be protected from animals by this uncanny and unusual behaviour.

But though the Sensitive Plant is certainly protected from grazing animals by these movements, other advantages may be derived. Heavy rain, for instance, such as occurs in the tropics, will not injure its delicate leaves. Dust-storms will not damage it, and at night there will be no loss of heat by radiation. The "shrunk" or folded condition of the leaflets will decrease any chance of injury by raindrops, for the rain will not fall on the broad surface of the leaflets. A nearly vertical leaf also will not suffer the loss of heat which a horizontal one would endure.

Besides the plants mentioned above, there are several others in which by a rather severe shaking the leaves can be made to fold up. This is the case with the common Woodsorrel (*Oxalis acetosella*), with the False Acacia (*Robinia*), and a few others.

The former has a peculiarly delicate leaf. In cold, wet weather its leaflets hang limp and numb from the leaf-stalk all day. In fine weather they are spread out horizontally. On a fine sunny afternoon its leaflets may sometimes take a mid-day sleep, for they hang loosely down in the same way that they do in cold, wet weather or at night.

But in the Woodsorrel these movements are not for protection against grazing animals.

There are other examples amongst plants of a distinct sudden movement which begins whenever part of the plant is touched. The movements of tendrils have been already referred to. The Venus' Fly Trap and the Sundew will be mentioned when we are discussing Insectivorous Plants. There are also several flowers in which the stamens suddenly spring up when they are touched by an insect (Barberry, Centaurea, and Sparmannia), and in Mimulus the style-flaps close when touched.

All these cases seem to involve some sort of mechanism which replaces the nervous system of animals.

No very definite laws have yet been discovered as to the way in which plants are affected by electricity, but enough is known to show that there are many interesting discoveries in prospect.

Professor Lemström has made some interesting experiments in the Polar regions which go to show that the rich development of plant life in that desolate region may be connected with the peculiar electrical conditions of the Polar atmosphere; the aurora borealis, which is a common phenomenon there, being also produced by those conditions.

Several writers have claimed that slight electric shocks given at frequent intervals help the growth of plants and especially quicken the germination of seeds, but it can scarcely be said that this has been proved.

When a branch or leaf-stalk is wounded or injured by being tightly clamped in a vice, then it will be found that a current of electricity passes from the injured spot to the part that is untouched, and then in the reverse direction.

Changes of current are also produced when a leaf is suddenly exposed to light for a short time and then shaded.

One of the most interesting observations is that made by Major Squiers near Lorin Station, in America, where the California Gas and Electric Corporation of San Francisco has a long-distance transmission telegraph line. The power is transmitted at a voltage of 56,000 with a frequency of sixty cycles per second (three-phase). Major Squiers, from previous experiments, thought that a note corresponding to this frequency might be heard in a telephone receiver. The following was the result: —

"Upon connecting the telephone between two nails driven in any growing tree along the route of the line, and at a reasonable distance therefrom, the telephone responded to this note with great clearness, and when the distance was not more than 100 feet, the sound was very loud. For this experiment no microphone need be used, nor any source of electromotive force other than that induced in the tree itself, the telephone being connected directly between two nails driven into the tree....

"Several kinds of trees of various sizes and forms were examined along this power transmission line, and all were found to be singing with a loud voice the fundamental note characteristic of the line current. Indeed, the strip of vegetation along this line has thus been singing continuously, day and night, for several years, since the operation of the line began; it needed only the electro-magnetic ear to make the sound apparent....

"The general appearance of vegetation along this route is certainly vigorous."

An interesting little experiment was carried out by the author in Glasgow, with the kind help of Professor Blyth, at the Glasgow and West of Scotland Technical College. By attaching one wire to the upper part of the stem of a young pot-plant whilst the other wire was inserted in the base of the stem, it was easy to show that an electric current was passing—at any rate, during the daytime. In the evening, however, this was not at all distinct. That such currents do occur in living trees seems to be admitted. A similar current was not found in a stick of dead-wood. The mere passage of the water through the plant in transpiration might, however, cause such a current, for the water is evaporated at the leaves.

A strong electric shock may of course *electrocute* a plant by killing the cells. It is possible to cause the Mimosa leaves to close by means of an electric shock.

CHAPTER XVI. ON FLOWERS OF THE WATER.

THE FIRST PLANT—SEAWEEDS IN HOT BATHS—BREAKING OF THE MERES—GORY DEW—PLANTS DRIVEN BACK TO THE WATER—MARSH PLANTS—FLEUR-DE-LIS—REEDS AND RUSHES—FLOATING ISLANDS—WATER-LILIES—*VICTORIA REGIA*—PLANTS 180 FEET DEEP—LIFE IN A POND, AS SEEN BY AN INHABITANT—FISH-FARMING—THE USEFUL DIATOM—WILLOWS AND ALDERS—POLLUTED STREAMS—THE HORNWORT—THE FLORIDA HYACINTH—REEDS AND GRASS-REEDS—THE RICHEST LANDS IN THE WORLD—PAPYRUS OF EGYPT—BIRDS AND HIPPOPOTAMI—FEVER AND AGUE.

WHAT was the first green plant? When was the surface of the earth first covered with flowers? Such questions are quite impossible to answer. We cannot even tell how plants ever came to exist on the earth at all. Wonderful as are the stories of the hardihood of bacteria, of spores, and of seeds, it is not possible to imagine that they could have been whirled or drifted through infinite space to this particular planet.

Yet it is at least probable that the first real plant on this world was a seaweed or alga.

In Germany and Austria there are certain springs in which the water coming from immense depths is at an exceedingly high temperature. These hot springs are used as natural hot baths, and have many interesting peculiarities. Amongst others there is the fact that certain seaweeds or algæ are found luxuriating in the hot water. Some of these can even live in springs with a temperature of 176° F.!

Such algæ may have remained living in exceedingly hot water ever since that long distant time, the very first of all the geological periods, when there was no distinct separation betwixt land and water, and when the waters which were below the firmament had not been separated from those which were above it. Then the world seems to have been all fog and mist at a very high temperature.

But all theories on the origin of the world might be briefly summarized by the last nine words!

At any rate, the first plant was almost certainly a seaweed or alga not unlike those which produce the so-called "breaking of the meres."

At some seasons the water of certain lakes, usually quite clear and pure, becomes discoloured, turbid, and everywhere crowded with multitudes of tiny, bright, verdigris-green specks. The fish at once begin to sulk, refuse to take the fly, and live torpid at the bottom of the water. The minute green particles consist of a certain seaweed or alga. Mr. Phillips put the head of a common pin in the water so as to obtain a very small drop. When placed under a microscope, this minute amount of water was found to contain 300 individual algæ.[0] This was in Newton Mere (Shropshire), and as this lake extends over 115 acres, it is possible to imagine the millions upon millions of algæ which must have existed in it. The names of these seaweeds are many thousand times longer than the algæ themselves, and it is not really necessary to give them. One of them, however, *Aphanizomenon flos-aquæ*, has been noticed "tingeing with its delicate green hue the margin of the smallest of the Lochs Maben, in Dumfriesshire." Yet it is not so big as the dot on the *i* in its name. Many other cases have been recorded of lakes that were coloured sometimes a "pea-green," or even brown or red by similar tiny little seaweeds. As we shall see, the water of such lakes generally contains a very large amount of suspended or floating vegetable life.

Another curious appearance is *Gory Dew*. Patches of a deep blood-red or purple colour are found on the ground or on walls. They have just the appearance of recently-shed blood. This also is due to an alga (*Porphyridium cruentum*). Dr. Cooke quotes from Drayton as follows: "In the plain, near Hastings, where the Norman William, after his victory found King Harold slain, he built Battle Abbey, which at last, as divers other monasteries, grew to a town enough populous. Thereabout is a place which, after rain, always looks red, which some have attributed to a very bloody sweat of the earth, as crying to Heaven for vengeance of so great a slaughter."

The ordinary "Rain of Blood" which appears on *not too fresh* meat, and looks like minute specks of red-currant jelly, is due to one of the Bacteria (*Micrococcus prodigiosus*).

The original algæ or seaweeds probably had descendants which migrated to the land and eventually after many geological periods became our flowering plants and ferns. But the earth has become so richly supplied with plants of all sorts and kinds that it is now by no means easy for any plant to find a roothold for its existence. So that a considerable number have been forced back to the water, and have accustomed themselves to live in or even under water in company with their lowly cousins, the seaweeds, who remained below its surface.

These water plants are very interesting. They are always competing with one another. There is a perpetual struggle going on round every pond and loch, and by every river side.

If you look carefully round the edge of a loch or pond which lies in a grass field, certain series of plants are generally found to follow one another in quite a definite way. The first sign of water in grass is generally the presence of moss or "fog" between the grass-stems and the appearance of what farmers call the "Blue Carnation Grass." It is not a grass but a sedge (*Carex glauca* or *C. panicea*) with leaves rather like those of a carnation. A little nearer the border of the pond, there may be a tall coarse grass (*Aira caespitosa* or *Festuca elatior*). Next there is almost certain to be a fringe of Rushes. Where the Rushes begin to find the ground too wet for them, all sorts of marsh plants flourish, such as Water Plantain, Cuckoo-flower, the Spearwort Buttercup, Woundwort, and the like. As soon as the actual water begins, one finds, whilst it is still shallow, the Flag series of yellow or purple Irises, Bogbeans, Marsh Cinquefoil, Mare's Tail, and Sedges of various kinds. In this part the water ranges from an inch or two to about eighteen inches deep.

The Flag or Iris is a very common and yet interesting plant. It has a stout, fleshy stem lying flat on the mud, and anchored to it by hundreds of little roots. The flower is the original of the Fleur-de-lis, or Lily of France, which took the fancy of the King of France as he rode through the marshes towards Paris. (It is true that there are some unromantic authors who hold that the emblem was really intended to represent a frog or toad!)

The flower consists of three upright petals and three hollow sepals, which make so many canals leading down to the honey, and roofed over by an arched and coloured style. As the bee hurries down the canal to its nectar, its back is first brushed by a narrow lip-like stigma and then dusted with pollen. The leaves overlap in a curious way, and, when they have withered, their stringy remains serve to protect the fleshy stem. *Orris root*, which is used in perfumery, is the stem of the *Iris florentina*.

Most of the other plants in this Flag series will be found to have prostrate main stems growing under the water, but giving off flowering and foliage stems which stand up above it, so that the leaves and flowers are above the surface.

In the next part of the pond, where the water is from eighteen inches to nine feet deep, masses of reeds will be found usually swaying, sighing, and whispering in the wind. There are many kinds, such as Bulrushes, Phragmites, Horsetail, Scirpus, etc. It seems to be the depth, the exposure to wind, the character of the soil, and other unknown factors, that determine which of those will be present. All of them are tall, standing well above the water; their main stem is usually flat on the bottom of the pond, or floating horizontally in the water, but giving off many upright branches.

Floating islands are often formed by some of these horizontal main stems breaking off and being carried away. Those Chinese who possess no land make floating islands of such reeds for themselves, and grow crops on them. There are hundreds of such islands in the Canton River.

A Leaf Raft

Victoria Regia, the giant water-lily of the Amazons. Those shown are in a public park in Minnesota and are able to support the weight of a child. In their native home the leaves are said to be ten feet in diameter.

Beyond the reeds, one sees the large flat, floating leaves and beautiful cup-like white or yellow flowers of the Water-lilies. They grow in water which is not more than fifteen feet deep. Their long stalks and leaf-stalks are flexible and yield readily, so as to keep the flowers and leaves floating. There are narrow submerged leaves as well. The actual stem of the White Water-lily is about three inches in diameter, and stout and fleshy. It is full of starchy material, and lies upon the mud deep down at the bottom of the pond. There are many advantages in the position of the flowers, for bees, flies, and other useful insects can reach them easily, but slugs, snails, and other enemies cannot do so. The little seeds have a curious lifebelt-like cup, which enables them to float on the surface.

Of course, our own British water-lilies cannot compare with the magnificent *Victoria regia* of the tropics. Its petals are white or pink on the inside, and its gigantic leaves, six feet or more in diameter, can support a retriever dog or a child. There used to be some of them at Kew Gardens. A curious point about these enormous floating leaves is that they are covered with little spiny points on the under side and at the margin; that is probably to keep some sort of fish from nibbling at the edges.

But to return to our pond. Beyond the water-lily region and so long as the water is from twelve to twenty-four feet deep, Pondweeds are able to grow, and their leaves may be seen in the water, whilst their stalks stand up above the surface so as to allow wind to scatter the pollen.

This depth of twenty-four feet seems at first sight very great, but it is a mere nothing compared with the regions entirely below the water, where certain Stoneworts (*Chara*) and Mosses have been found flourishing. The former has been dredged up from depths of ninety feet, and a little moss was discovered in the Lake of Geneva growing quite comfortably at a depth of 180 feet below the surface.

But it is quite impossible to appreciate the wonder and beauty of the life in a pond unless by a strong effort of the imagination.

Suppose yourself to be a fish two or three inches long, and accustomed to the dim, mysterious light which filters down through the water from the sky above. Every here and there great olive-brown leaf-stalks and stems cross and, branching, intercept the light. Everything, the surface of the mud, the stems and branches of the submerged water-plants, is covered by an exquisite golden-brown powder, which consists of hundreds and thousands of "Diatoms." Here and there from the Pondweed and other stems hang festoons or wreaths or threads of beautiful green Algæ. Little branching sprays of them, or perhaps of the brown kind, are attached here and there to the thick stems.

Even the very water is full of small, floating, vivid green stars or crescents or three-cornered pieces which are free floating Algæ or Desmids. Other diatoms are also free or swim with a cork-screwing motion through the water. Great snails and slugs crawl upon the plants, and weird large-eyed creatures, with a superfluity of legs and an entire absence of reserve as to what is going on inside their bodies, skirmish around. So that such a pond is full of vegetable activity. The free-swimming diatoms and desmids make up the food of the snails and crustaceans. These latter in turn are the food of fishes.

It is even possible to-day by carefully stocking an artificial pond with water plants, by then introducing Mollusca and Crustacea, and finally by the introduction of "eyed ova" or fry of the trout, carp, or other fishes, to produce a regular population of fishes which can be made more or less profitable, and the process can be spoken of as "fish-farming." Unfortunately there are a great many gaps in our knowledge as to what fish actually feed on, and we know even less about what the Mollusca and Crustacea require.

There is, however, a distinct annual harvest of these minute seaweeds, of which different sorts appear to develop one after the other, just as flowering plants do. The two months January and February, which are almost without flowers, are also those in which most of these minute vegetables take their repose in the form of cysts or spores.

But these diatoms are too important and too interesting to be dismissed in such a cursory manner. Each consists of a tiny speck of living matter with a drop or two of oil enclosed in a variously sculptured flinty shell. They have, in fact, been compared to little protected cruisers which pass to and fro in the water and multiply with the most extraordinary rapidity.

If you (1) use dynamite to blast a rock, (2) if you employ a microscope or telescope, (3) if you paint an oil picture, (4) if you make a sound-proof partition in a set of offices, the probability is that it has been necessary to use the substance diatomite in each case. This consists of the accumulated shells of myriads of diatoms.

Nor does that represent by any means the whole of the usefulness of these tiny seaweeds. The oil shales, such as occur in Linlithgowshire and elsewhere, are supposed to be the muddy, oily deposits of such ponds as we have endeavoured to describe. The oil found in the shales was probably worked up by these diatoms in long-past geological ages. It may be used to-day either (1) to drive motors, (2) to light lamps, (3) to burn as so-called "wax" candles, (4) to eat (as an inferior sort of chocolate cream).

Interesting as these diatoms are, it is not really possible to understand their structure without the use of a microscope, so that we must pass on to another side of the activity of water plants.

Let us, for instance, notice some of the ordinary plants to be found along a riverside. Willows and Alders are the ordinary trees, because they are specially fitted to stand the danger of being regularly overflowed. They easily take root, so that branches broken off and floated down are enabled to form new trees without much difficulty. In the United States, it has become a custom to plant Willows along the banks, because they are then not so liable to be broken down and worn away. Yet when a big Willow tree has become undermined, the weight of the trunk may cause it to fall over towards the water, so that a large section of the bank may be loosened and serious damage may be done if it is torn away by a heavy flood.

Amongst such Willows, should be mentioned the "cricket bat" kind, which has to be grown with the very greatest care, and of which a single tree may be worth £28.

Many of our rivers are, alas, sadly polluted by artificial and other impurities which kill the fishes and destroy the natural vegetation. When this happens a horrible-looking whitish fungus (*Apodytes lactea*) coats the stones and banks under water and the water swarms with bacteria. This fungus and the bacteria are really purifying the water, for they break up the decaying matter in it.

The oily or slimy character of the outside skin of all submerged plants is of very great importance to them. It allows the water to glide or slip over them without any friction.

Still keeping to our river bank, let us look for submerged plants. What is that dark green feathery plume? It is the Hornwort (*Ceratophyllum*) gently wriggling or moving from side to side. It has probably never been still for a moment since it first began to grow. Take it out of the water, and it collapses into a moist, unpleasant little body, but as soon as it is put in its natural element again it is seen to have a thin flexible stem along which there are circles of curved, finely divided leaves. Watch it in the water and one is filled with astonishment at the perfection of the shape, arrangement, and character of the leaves, which enables them to hold their place even when a flood may cover them with an extra twenty feet of water! The same sort of leaf, but with great difference in detail, is found in the submerged Water Crowfoot, Water Milfoil, Potamogetons, and others which live under the same conditions.

If it were the St. John's River, we might see that extraordinary Florida Hyacinth which has swollen, gouty-looking leaf-stalks, and grows with such extraordinary rapidity that it covers the whole surface of rivers, choking the paddle-wheels of steamers and destroying the trade in timber, for no logs can be floated down when it covers the water. Its rosettes float on the surface, and are very interesting to examine. If you upset one or turn it upside down in the water, the "buoys" or swollen stalks act as a self-righting arrangement, and it slowly returns to its proper position.

But in most rivers, one is certain to come across backwaters where it is impossible to force a boat through on account of the reeds and other marsh-plants.

There are places on the Danube where hundreds of square miles are occupied by waving masses of the feathery-plumed Phragmites, almost to the exclusion of any other sort of vegetation. Giant specimens of it eighteen feet high have been observed.

The same reed occurs in North and South America and far up towards the Arctic regions. At first sight it seems as if this was a mistake of Nature; why should so much of the surface be occupied by this useless vegetable? But it is necessary to say a little more about its habits and its object in life.

The most interesting and curious point is the way in which it grows in dense thickets; the main stem is really horizontal and below the water, but it gives off a number of upright stalks. Now every flood will carry in amongst the stalks quantities of silt and rubbish. Those upright stems will sift the water: all sorts of floating material, sand, silt, dead leaves, fruit, etc., are left amongst them. So that such a marsh or bed of Phragmites is gradually, flood by flood, collecting the deposits of mud, and the bed becomes every year more shallow. At the edge of the marsh there is scarcely any water visible, and grasses and other plants are beginning to grow between the Phragmites stems. Eventually these latter are choked out, and a marshy alluvial flat occupies the site of the old reed-bed.

So that the work of Phragmites is of the greatest possible importance: it has to form those fertile alluvial flats which are found along the course of every great river, and which are by far the most valuable lands in the whole world.

Look, for instance, at the population of Belgium, Holland, and Lower Germany, and notice how dense it is upon the alluvial flats where the Meuse, Rhine, and other rivers approach the sea. It is just the same in Britain. London lies on the great alluvial flats of the Thames, Glasgow on the Clyde, Liverpool on the Mersey. In China it is the Yang-tze-kiang valley (especially near its mouth); in India, the Ganges, of lower Bengal, and in the Argentine the La Plata River, which show the greatest accumulations of humanity. In every case it is the rich flat alluvium, which is exceedingly fertile when drained and cultivated, that has originally attracted so many people.

Lower Egypt is the gift of the Nile, but it is not so much the Nile as these neglected water plants which made the rich lucerne, cotton, and food crops of Lower Egypt possible. Amongst the Egyptian Reeds one especially is of great importance. The *Papyrus antiquorum*, ten feet high, has much the same habit as our Phragmites and other water plants. It forms dense, almost impassable thickets, sometimes completely occupying and choking a small valley, or leaving only a passage, often changing and half choked, through a larger one. This, with other plants, makes the "sudd" of the Nile, which is one enormous accumulation of marsh plants and reeds floating on the water and covering a length of over 500 miles.

It was from the Papyrus that the ancient Egyptians made their paper. The stems are six to seven inches in diameter. "The pith of the larger flowering stems ... cut into thin strips, united together by narrowly overlapping margins, and then crossed under pressure by a similar arrangement of strips at right angles, constitutes the Papyrus of antiquity."

These great marshes and reed-beds are full of interest to naturalists. The Fens of Lincolnshire and the Norfolk Broads show the way in which water plants keep hold of the worn and travelled rubbish of the hills, and prevent most of it from becoming useless, barren sea-sands. These places, however, like the sudd of the Nile, and the Roman "Campagna," have an evil reputation so far as climate is concerned. This used to be the case even in lower Chelsea, in London (where snipe were shot not so very long ago). It is as if Nature had desired to do her own work in peace and without being disturbed, for fever, ague, mosquitoes, and malaria are very common. Yet a certain number of people always live in such places. In France, e.g., the leeches in the great marshes near the Landes form a source of riches. Such reeds also are or were the home of the hippopotamus, crocodile, and other extraordinary animals. The extinct British hippopotamus no doubt found in the Chelsea or other marshes a home as congenial to its tastes as is the sudd of Egypt to its living descendants or allies. In other places the enormous quantities of water birds, myriads of ducks, geese, swans, regiments of flamingoes, snipe, and the like, have called into existence peculiar kinds of industry in fowling and netting that are not without importance. The decoys in the Fens yield hundreds of birds for the London market, and the duck-punts with their huge guns also bring in quantities of wild fowl.

But all this industry is very trifling compared with that of Phragmites and its associates, who have strained from the water of the Thames most of the ground on which London now stands.

CHAPTER XVII. ON GRASSLANDS.

WHERE IS PEACE?—TROUBLES OF THE GRASS—ROADSIDES—GLACIERS IN SWITZERLAND—STRENGTH AND GRACEFULNESS OF GRASSES—RAINSTORMS—DANGERS OF DROUGHT AND OF SWAMPING—ARTIFICIAL FIELDS—FARMER'S ABSTRUSE CALCULATIONS—GRASS MIXTURES—TENNIS LAWNS—THE INVASION OF FOREST—NATURAL GRASS—PRAIRIE OF THE UNITED STATES, RED INDIAN, COWBOY—PAMPAS AND GAUCHO—THISTLES AND TALL STORIES—SOUTH AFRICA AND BOERS—HUNTING OF THE TARTARS—AN UNFORTUNATE CHINESE PRINCESS—AUSTRALIAN SHEPHERDS.

WHERE should one seek for peace on earth? The ideal chosen for one well-known picture is a grassy down "close clipt by nibbling sheep," such as the fresh green turf of the South Downs.

Others might prefer the "Constable country," near perhaps the famous "Valley Farm" of which the picture now hangs in the National Gallery, and especially in early spring. At any rate, once seen, one remembers for ever afterwards those glossy-coated, well-fed, leisurely cows grazing hock-deep in rich meadows full of bright flowers and graceful grasses, through which there winds a very lazy river bordered by trim pollarded willows.

The charm of the South Downs and of Constable's meadows depends upon their peaceful quiet, and the absence of any sign of the handiwork of disturbing man.

But such meadows are entirely artificial. They could no more exist in nature than a coal-mine, if it were not for man's help. Moreover, they are in a state of perpetual war! No plant within them experiences the blessings of peace from the time it germinates until the day that it dies.

Each plant is fighting with its neighbours for light, for air, for water, and for salts in the soil, and it is also trying to protect itself against grazing animals, against the vole which gnaws its roots, and against the insects and caterpillars which try to devour its buds.

Besides its own private and individual troubles, it is but one of a whole company or army of plants which, like a cooperative society, occupy the field.

Other societies, such as peat-moss, thickets, and woods, try to drive out the grasses and cover that particular place in its stead. The Grassland companions are also always trying to take up new ground, and to cover over any which is not strongly held by other plants.

A road, for instance, is always being attacked by the grassland near it. It is sure to have a distinct border of Rat's Tail Plantain, Dandelion, Creeping Buttercup, and Yellow Clovers. These are the advanced guard of the grassland. However heavily you tread upon these plants, you will do them no injury whatever, for they are specially designed to resist heavy weights. But, if the road were only left alone, these bordering plants would be very soon choked out. The ordinary buttercup would replace the creeping species, and white or red clovers take the place of the little yellow ones, whilst grasses would very soon spring up all over it.

But of course the roadman comes and scrapes off all the new growth of colonizing grasses, etc. Then the plantains, dandelions, and yellow clovers patiently begin their work again.

These sequoias grow to from 250 to 400 feet high, though they are not quite the tallest trees in the world.

In Switzerland, in those valleys in which the glaciers are melting away, leaving stretches of bare mud, scratched stones, and polished rock, plants immediately begin to settle there. A Swiss botanist watched the process during five or six years, and describes how first the yellow Saxifrage (*S. aizoides*) establishes itself. Next season Coltsfoot, willow-herb, Oxyria, and two grasses had planted themselves. During the third season another grass came in. By the fourth season, Fescues and yarrow had appeared, and by the fifth season, five grasses, clovers, and yarrow had formed a regular grassland upon the new untouched soil.

In such cases, Nature, who abhors bare ground, is endeavouring to clothe it with useful vegetation.

The fights which are going on are of the most ruthless character. Many weeds are said to produce some 30,000 seeds in one year, and every plant which grows in a meadow is scattering thousands of seeds. But of course the number of plants remains much the same, so that 29,999 seeds are wasted (or the seedlings choked out) for every one that grows up!

It is probably because of this perpetual warfare that the growth of the grasses is so vigorous, and their whole structure so perfectly adapted. If you watch a flowering grass, you are sure to notice how narrow is its stem compared with the height. A factory chimney only fifty-eight feet high requires to be at least four feet broad at the base, yet a ryeplant 1500 millimetres high may be only three millimetres broad near the root. Man's handiwork, the chimney, is in height seventeen times its diameter, but the height of the grass is 500 times its diameter.

The neatness of design, the graceful curves and perfect balance in the little flowering branches at the top of a haulm, is always worth looking at, and particularly in the early morning when it is beset with sparkling drops of dew.

It is all wiry, bending and swaying to the wind so as to produce those waves which roll across a hay-field, and on which the shimmering light is reflected and changes colour. The fight for light and air, the struggle to get their heads up above their competitors, produces all this exquisite mechanism.

It is true that a heavy rainstorm may beat the stems flat down to the ground, but, as soon as the weather becomes dry again these same stems will raise themselves up and become upright; they have a special sensitiveness and a special kind of growth which enables them to do this.

There are two special dangers which all such artificial meadows have to withstand. Let us see what will happen if such a meadow begins to dry up through a sinking of the level of the water below the soil.

Each grass has its own special favourite amount of moisture. It likes to have its water at just one particular depth below the surface. Unfortunately there are not nearly enough sympathetic and careful observations of the preferences of each individual grass. A Danish author has worked out the facts in certain localities (Geest). Suppose first that the water-level of the wells, etc., is 6-1/2 to 9-3/4 feet below the surface. This suits the Meadow Poa grass (*Poa pratensis*) exactly. It will grow luxuriantly and flourish. Now suppose the weather is very wet, so that the water rises in the wells till they are three to four feet deep. The Roughish Poa (*P. trivialis*) prefers this moister soil, and it will grow so vigorously that it will kill out the other kind. If it is a season of very heavy floods, or if the drains become choked so that the water rises to within fourteen to twenty-five inches of the surface, then the tufted Aira (*Deschampsia caespitosa*) will kill out the other kinds and flourish abundantly. But if the water rises higher than this the marsh series comes in (see Chap. XVI.).

So that the thirsty grasses of the meadow are helped or hindered in their fight for life by changes in the water away down in the soil below their roots.

Even in Great Britain one can see distinct differences in very dry and very wet summers, but all these pastures, meadowlands, and hay-fields are, as we have already mentioned, as much due to man's forethought and industry as a factory or coal-mine.

It is very difficult to realize this. The best way is to go to the National, or any other good picture-gallery, and look carefully at any landscapes painted before the year 1805. You will scarcely believe that the country as painted can be the land we know. Where is the "awful orderliness" of England? Where are the trim hedges? Where are the tidy roadsides and beautifully embanked rivers that we see to-day?

As a matter of fact, until the great Macadam made good roads and the great Telford and other engineers built stone bridges, it was impossible to rely on getting about with carts and carriages. Gentlemen's coaches and wagons used to be literally stuck in the mud! Horses were drowned at fords, or died in their struggles to pull very light loads through mud which nearly reached the axles of the wheels (see Chap. XI.).

Besides the change due to roads, fences, drains, and farm buildings, the very grasses themselves are growing unnaturally. The farmer has selected and sown what he thinks best.

He is obliged to do so, because grasses vary so much. Some of them shoot up quickly and die after the first year. Others live for two years, whilst a great many bide their time, developing very slowly, and not reaching their full growth until the fourth or fifth year.

Some are tall and vigorous, others are short; some flower early in the season, and others very late. Many send out quantities of suckers or runners at the base, so that they form a dense, intricate turf—a mass of stems and roots thickly covering the ground.

A farmer wants his pasture to begin early and to continue late; he must have a good first year's crop, and it must remain good for years afterwards. So that his calculations as regards the proportions of the different grass seeds which he requires are of the most abstruse character.

To sow such "permanent pasture," prepared by blending together grasses and clovers with an eye to all the above necessities, there will be needed some seven million seeds for every acre.

The art consists in coaxing the good, lasting, nutritious ones to make both tall hay, rich aftermath, and a close, thick turf below, and, until these are ready, to use the annual and biennial grasses.

Such beautifully shaven, green, soft turf as one sees in the lawns of cathedrals or the "quads" at Oxford and Cambridge has been most carefully and regularly watered, rolled, and mown for hundreds of years. It is not easy to keep even a tennis-lawn in good condition. Little tufts of daisies appear. Their leaves lie so flat that they escape the teeth of the mower, and they are not so liable to be injured by tennis-shoes as the tiny upright grass-shoots which are trying to spring up everywhere. The Plantain is even9 worse, for it is specially built to stand heavy weights, and it has several roots which divide and branch like the prongs which fix teeth in the jaw, so that it is very difficult to howk it out.

Thus our grasslands in Britain are unnatural and artificial productions. If the field drains are choked, moss or fog and rushes appear. Still more interesting, however, is what happens if the farmer is not careful to destroy the taller weeds, such as Dock, Ragweed, Cow Parsnip, Thistles, and the like. If you walk over a grass-field in early spring, you are sure to see some of these pests. At this stage they have a very humble, weak, and innocent appearance: they are quite small rosettes or tufts. Yet they are crowded with leaves, which are hard at work busily manufacturing food material. Soon they begin to shoot up. Their leaves overreach all the neighbouring grasses. Their roots spread in every direction, taking what ought to go to the "good green herb intended for the service of man." They finally accomplish their wickedness by producing thousands of seeds, which are scattered broadcast over the fields.

By this time the farmer sees what is going on, and endeavours to cut them down; but it is a long, slow, and laborious proceeding. One year's seeding means seven years' weeding.

Yet these tall Thistles and Ragweeds are only the first stage of a very interesting invasion. Look around the field corners, on railway-banks, or in old quarries, where man has left things alone. You will see these same tall herbs (the Ragweed, etc.), but you are sure to find a place where they are being suppressed by Rasps, Briers, and Brambles. These are taller, stronger, and more vigorous than the herbs, and they also last longer, for their leaves are still at work in November. This is the second stage of the invasion. But if the place has been long neglected, Hawthorns and Rowans, Birch and Ash will be found growing up. These last show what is happening.

A wood is trying to grow up on the grassland. If left alone, an oak or beech forest would, after many years, spread over all our grass pastures and hay-fields. These tall herbs are the pioneers, and the briers and brambles are its advanced guard.

As a matter of fact, by far the greatest part of our agricultural land *was* a forest, but it has been cut down, drained, dug, weeded, hedged, and "huzzed and maazed" with agricultural implements and more or less scientifically selected manures, until it is made to yield good beef, excellent mutton, and almost the largest crops per acre in the world.

Natural grasslands exist, however, in every continent.

The great Steppes of Southern Russia and the pastures that extend far to the eastward even to the very borders of China, the Prairies of North America, the Pampas of Argentina, the great sheep-farms of Australia, and a large proportion of South Africa, consist of wide, treeless, grassy plains, where forests only occur along the banks of rivers, in narrow hill-valleys, or upon mountains of considerable altitude. Upon these great plateaux or undulating hills the rainfall, though it is but small in amount, is equally distributed, so that there is no lengthy and arid dry season. Take the American Prairie, for instance. These valuable lands, once the home of unnumbered bison and hordes of antelopes, lie between the ancient forests of the eastern states and the half-deserts and true salt deserts of the extreme west. Rivers, accompanied in their windings by riverside forests, are found (especially in the east). The real prairie has a blackish, loamy soil, covered sometimes by the rich Buffalo or Mesquite grass, which forms a short, velvety covering, not exactly a turf such as we find in England, but still true grassland. It is only green in early spring.

The Bushman is levering up the root of elephant's foot to get the starchy food inside. He does it by a stick run through a rounded stone. The woman has caught a lizard for the boy to eat.

From the spring onwards until the end of summer there is an endless succession of flowers. The first spring blossoms appear in April; great stretches are covered with Pentstemons, Cypripediums, and many others in May and June; then follow tall, herbaceous Phloxes, Lilies, and Asclepiads, but perhaps the most characteristic flora blossoms still later on, when every one "wants to be in Kansas when the Sunflowers bloom." Over these prairies used to travel the great wagons or "prairie schooners." The cowboy, who almost lives on horseback, watches over great herds of cattle and troops of half-wild horses. Yet his life is, or used to be, almost as free, comfortless, and uncivilized as that of the buffalo-hunting Indian who preceded him. One must not forget to mention the prairie-dog—able to utilize the abundant grass, and diving into a safe refuge underground when threatened by the wolves or other carnivorous creatures, which, of course, multiplied exceedingly, thanks to the jack-hare, antelopes, and bisons.

The Pampas in South America is a similar grassland. On the east it stops at the woodlands along the great Plate River, but on the west it becomes gradually more dry and arid, until long before the Andes are reached it is too dry even to carry sheep, and can only be described as a half-desert.

"It is a boundless sea of grasses fading into the distant horizon, which can only be distinguished when the sun is rising or setting." Yet amongst the grasses are hundreds of flowers, and, a fact which is very remarkable, many of them, such as Fennel, Artichoke, Milk Thistle, Burdock, Rye Grass, etc., are European plants which have dispossessed the natives over miles of country, exactly as the gaucho has driven away or exterminated the Indians who lived there. It is covered by tufts of grass betwixt which appears the rich alluvial earth, yet in good years it may become almost a perfect grass floor. "The colour changes greatly, for in spring when the old grass is burnt off, it is coal-black, which changes to a bright blue-green as soon as the young leaves appear; later on it becomes brownish green, which again changes when the silver-white flowers come out to the appearance of a rolling, waving sea of shining silver."

Here would be the place to mention how an army encamped upon the Pampas finds itself next morning imprisoned and doomed to perish miserably in a forest of giant thistles which has sprung up during the night. There is no doubt that thistles and other weeds are very tall in both South and North America. Fennels are ten to twelve feet high, and even little Chenopodiums (such as in England may reach eighteen inches), become in South America seven to eight feet high, but the tallness of some of the stories is more remarkable even than that of the plants!

Over the Pampas used to roam thousands of guanacos (a creature of the most unlovely type, which resembles both a camel, a mule, a deer, and a horse); here also were Darwin's ostriches (*Rhea Darwinii*) and other game, which were caught by the lasso and by the peculiar "bolas" of the Indians. They used to surround the herds and then massacre them by hundreds. The "tuco tuco" also, which is a burrowing rodent with habits very like those of the prairie dog, finds plenty of sustenance in the abundant grasses. Upon them subsist pumas, foxes, and other carnivores.

We have said that the Pampas gradually changes from being very fertile on the east to being almost a desert on the west. Here is the place to mention a very interesting, if not romantic, fact. The guanaco does *not* travel hundreds of miles in order to die in one particular spot as soon as it feels ill, but it does resort especially to certain spots. There the grass is often a bright, fresh green, for it is plentifully manured, and consequently the guanaco helps to encourage the good grasses to occupy a half-desert. On the eastern side of the Pampas great changes are beginning to appear. The owners of the great camps, haciendas or cattle-ranches let off small parts of their land to Italian "colonists." These people grow crops of Indian corn, and when that has been reaped, the valuable Alfalfa or Lucerne is sown down. This forms the most exquisite and valuable pasture, and consequently far more Shorthorn and Durham cattle can be maintained.

There are in South Africa enormous grassy plains, where once springbok and other game used to exist in enormous herds (Wangeman records having seen a herd of antelope four miles long), in spite of lions and other beasts of prey, and in spite also of the Boer, who was as much a horseman as the gaucho or Red Indian. The great buck wagons of South Africa were almost as much the real homes of the Boers as the two-roomed huts which make up his "farms."

The great Steppes of Russia and Siberia are also grasslands. "As seen from a distance hills covered by the Stipa grass resemble sand-hills, but, when nearer at hand, the sand-grey colour changes into a silvery white, and these ever-moving grasses remind one of the waves of the ocean and, in spite of their monotony, leave a pleasant impression."

Tulips, Hyacinths, Veronicas, Periwinkles, Scotch Thistles, Euphorbias, Wormwoods, and other of our common plants or their near cousins, make up most of the flora of the Steppes. Yet there are hundreds of others, for it is a vegetation very rich in species.

If one reads in Gibbon's stately language of the mode of life of the Huns, the Scythians, and those other barbarians who, originating in these huge grasslands, occasionally overflowed and overwhelmed the civilization of declining Rome, the resemblance to Red Indians, Pampas Indians, cowboys, gauchos, and Boers is not a little striking.

Read, for instance, the magnificent account of the great hunting matches of the Tartar princes. "A circle is drawn of many miles in circumference, to encompass the game of an extensive district; and the troops that form the circle regularly advance towards a common centre, where the animals, surrounded on every side, are abandoned to the darts of the hunters." Both the Red Indians of the Prairie and the savages of the Pampas used to surround and destroy the game in exactly the same way.

The unfortunate Chinese princess given over for political advantages to a prince of the Huns, "laments that she had been condemned by her parents to a distant exile, under a barbarian husband, and complains that sour milk was her only drink, raw flesh her only food, a tent her only palace." This describes exactly the ordinary life and home of the Huns. "The Scythians of every age have been celebrated as bold and skilful riders; and constant practice had seated them so firmly on horseback, that they were supposed by strangers to perform the ordinary duties of civil life—to eat, to drink, and even to sleep—without dismounting from their steeds." Red Indians of Pampas and Prairie, cowboy and gaucho, lived exactly in the same way.

In those pages of Gibbon which treat of the Huns, Scythians, and other hordes, one recognizes sometimes the wagon of the Boers; sometimes a migration of the East African Masai; then perhaps it is a weapon that is really the lasso, or a disposition and character exactly paralleled by the Crows and Blackfeet. Even the great grass plains of Australia, where the kangaroo, the wallaby, and the dingo have been replaced by the sheep and the "Waler" horse, one finds, in the shepherd and squatter, traits that remind one of the gaucho or the cowboy.

Nor is this in the least extraordinary, for when a scanty rainfall produces those great limitless rolling seas of grass, Nature provides first large herbivorous animals to eat it down as well as carnivorous beasts to keep their numbers in control, until such time as a race of horsemen appears, whose domestic cattle replace the bisons, guanacos, kangaroos, and antelopes, and so assist in replenishing and subduing the earth.

CHAPTER XVIII. POISONS.

POISONED ARROWS – FISH POISONS – MANCHINEEL – CURARE – A WONDERFUL STORY – ANTIARIS – ORDEALS – THE OBI POISON – ORACLES PRODUCED BY POISONS – PLANTS WHICH MAKE HORSES CRAZY AND OTHERS THAT REMOVE THEIR HAIR – AUSTRALIAN SHEEP AND THE CAUSTIC CREEPER – SWELLED HEAD – MADNESS BY THE DARLING PEA – WILD AND TAME ANIMALS, HOW THEY KNOW POISONS – HOW DO THEY TELL ONE ANOTHER? – THE YEW TREE, WHEN IS IT, AND WHEN IS IT NOT POISONOUS?

EVEN to-day all embryo chemists and doctors are required to "pass" in the recognition of the more important medicinal plants.

But their knowledge is probably very superficial as compared with that of a bushman in the Kalahari Desert of South Africa. Every man, woman, and child in such a tribe knows thoroughly every plant that grows in the neighbourhood. His diet is a varied one, for it includes maggots, fish, frogs, snakes, white ants, and other horrible ingredients, but he lives mainly on roots, bulbs, and herbs of sorts. In times of famine he has had to obtain the most intimate knowledge possible of many plants, that namely which is obtained by eating them, and he has most carefully observed the poisonous kinds. These latter have given him, too, a very powerful weapon, for it is the poisoned arrows which give him the chance of killing game, otherwise utterly beyond his reach. He is on the fair road to becoming a hunter and tribesman, instead of being only a member of a morose, outcast family, always wandering and always hungry.

Probably poisons were first used in fishing. Many vegetable drugs, when thrown into pools and lakes, have the property of stupefying or killing the fish. A great many of these fish poisons are known, and it is quite easy to use them.

Amongst the Dyaks of Borneo, screens of basketwork are placed along a stream to prevent the fish escaping. Then the Dyaks collect along either bank in their canoes. Everybody has a supply of the root of the tubai (*Menispermum sp.*), which they hammer with stones in the water inside the canoe, so as to extract the poison. At a given signal the poisonous stuff is baled into the river, and very soon afterwards a scene of wild excitement begins, for the fish are speared or captured with handnets as they rise, stupefied, to the surface. The women scoop up the small fry in their nets.

Even at the Sea of Galilee, Tristram mentions that Arabs sometimes obtain their fish by poisoned bread-crumbs. In the South Sea Islands, at Tahiti, a poison is obtained from the nuts of a kind of Betonica, and is used to catch the fish among the reefs near shore. In West Africa several fish poisons are in use (e.g. seeds of *Tephrosia Vogelii*), and probably the same methods are used almost everywhere. They are by no means extinct even at home, for the occasional poacher sometimes uses fish poisons.

Arrow poison is, however, much more important, and is used by a great number of tribes in almost every part of the world. In 1859, in a war with the Dyaks of Borneo, the English army lost thirty men by poisoned arrows. They are deadly weapons, for the dart is a very thin piece of reed or cane, which has been dipped in the Upas poison (*Antiaris toxicaria*). It is propelled from a blow pipe, which in practised hands is able to carry 250 feet. One or two of these darts may cause death in two hours' time. The Spaniards, in their conquest of the West Indian islands, were often defeated by the poisoned arrows of the Caribs. The wounded died in agonies of suffering and delirium, sometimes protracted for twenty-four hours after receiving the wound.

The poison in this case is supposed to have been the Manchineel (*Hippomane*).

It is a handsome tree, but a very dangerous one, for the slightest cut on the surface produces a flow of a very fine white milk which is acrid and poisonous. This juice produces temporary or total blindness if the slightest speck enters the eyes, or even if one sits over a fire made of its wood. It is probably not true that people are killed if they merely sleep below it, and grass will probably grow quite well under its shade, although there are stories which deny this. Blowpipes and poisoned darts are used by many savages in Asia and South America. Perhaps the Curare or Woorali poison is the most wonderful of the South American kinds. The tree, *Strychnos sp.*, grows along the Amazon and in the Guianas. The poison is obtained from the wood and bark, and several other vegetable substances are mixed with it. (This is a very common feature of native drugs and increases the chances of doing *something*.) It is a blood poison, and a very deadly one. Large animals like the tapir stagger about, collapse, and die after a very few steps, if they have been wounded by a dart. Humboldt declares that the earth-eating Otomaks were able to kill their antagonists by the mere pressure of their poisoned thumbnails.

In Africa it is more usual to find poisoned arrows shot from a bow. The exquisitely beautiful seed of *Strophanthus Kombe* is used as an arrow poison. The plant is a climber found in forests or bush, and has large woody pods about seven to twelve inches long. When these are open, the inside is seen to be full of the small yellowish seeds; each ends in a fine awn three to four inches long, which carries at the end a beautiful tuft of the finest silky hairs. The seed-coat is also covered with silk hairs. When viewed against a black surface, there is no more lovely object in nature. Yet from the seed-coat a very deadly poison is obtained; probably snake-venom and various gluey substances form part of the mixture, which is daubed on the arrows. Dr. Kolbe saw the Hottentots plastering their arrows with the poison of the hooded snake. Bushmen use a Lily bulb, *Haemanthus toxicarius*, but sometimes add part of the inside of a small caterpillar.

Another African poison which is not so well known is the *Acokanthera*, which was the ingredient in the arrows obtained by the writer in British East Africa.

North America is singularly free from these unsportsmanlike and horrible weapons, but they were not unknown in Europe in very ancient times. Pliny speaks of the Arabian pirates as poisoners, and allusions to their use of deadly arrows can be found in Horace, Ovid, and Homer. In the *Odyssey*, the hero goes to Ephyra (Epirus?) to purchase a deadly arrow poison, but he is refused for fear of the eternal gods. Poisoned arrows were employed by the Celts in Gaul, and also by the Saracens in the War of Granada in 1484.

Yet even in the time of Homer the sense of humanity seems to have decided against poisoned arrows as being both unnecessary and cruel, just as, in our own times, explosive bullets have been condemned, and are no longer used by civilized nations. But we should remember that until man became so expert with the bow and spear and so civilized by tribal fights as to be able to do without poisons, they were a very useful help in the struggle for civilization. Hundreds of thin pieces of bamboo about six inches long were regularly carried by certain African tribes. When dipped in poison and afterwards placed in paths in the ground, they formed a very efficient protection against barefooted enemies.

The Antiaris alluded to above is the famous Upas tree of Java. The tree was *said* to grow in a desert with not another living plant within ten miles of it. Such was the virulence of its poison that there were no fish in the waters. Neither rat, nor mouse, nor any other vermin had ever been seen there; and when any birds flew so near this tree that the effluvia reached them, they fell dead—a sacrifice to the effects of its poison. Out of a population of sixteen hundred persons who were compelled, on account of civil dissensions, to reside within twelve or fourteen miles of the tree, not more than three hundred remained alive in two months. Criminals condemned to die were offered the chance of life if they would go to the Upas tree and collect some of the poison. They were provided with masks (not unlike our modern motor-veils), and yet not two in twenty returned from the expedition.

All the foregoing statements were for years implicitly believed. They were vouched for by a Dutch surgeon resident in Java. Medicine is a profession, and Holland is a country which would in no way lead one to expect such magnificent mendacious audacity!

For the whole of the preceding statements about Antiaris is pure romance. The inner bark of young trees, when made into coarse garments, produces an extremely painful itching, whilst the dried juice is a virulent arrow poison.

Hellebore and Aconite were the favourite poisons of the Marquise de Brinvilliers and other specialists of the Middle Ages. The Christmas Roses or Hellebores were known to be poisonous fourteen hundred years before the Christian era, and are still used in medicine. Aconite, which has a tuberous root-stock, is dangerous, for it is occasionally eaten in mistake for the horse-radish, to which it has a faint resemblance. All kinds of aconite are poisonous. That of one of the Indian species is used to tip the arrows employed in shooting tigers.

Trials by ordeal were very common in ancient times. The theory was that an innocent person was not injured by certain drugs, which, however, proved immediately fatal to the guilty.

Such trials at one time were customary in almost every part of the world. They were supposed to be perfectly just, so that no man could be held guilty of the death of those who succumbed. In practice, however, they were almost invariably corrupt. The *Tanghinia venenifera* of Madagascar was regularly used in ordeals, and is probably still employed by certain tribes. The seeds are exceedingly poisonous, but, if the authorities wish the accused person to escape, a strong emetic is mixed with the powdered seeds, and the poison has no time to act. This, however, is seldom the case, for in any savage nation no one who is popular and in good esteem with the king or other people in authority is at all likely to be accused. The fact of his being accused means in most cases that he is already condemned to die. Another ordeal plant is the Calabar Bean (*Physostigma venenosa*), found in West Africa. The plant is a climber belonging to the *Leguminosæ*, and the seeds, which are about an inch in diameter, are very deadly. The seed is conspicuously marked by the long, dark, sunken scar, where it was attached to the pod. Besides being exceedingly poisonous, it has also a curious effect upon the pupil of the eye, which is contracted by this drug.

Another famous poison is produced from *Datura stramonium* and allied species. In tropical and sub-tropical countries, one is almost sure to find specimens of this handsome plant along almost every roadside. It is in fact one of the commonest tropical weeds. The leaves are large with fine spinose margins, and the flower is most conspicuous, as it is four or five inches long. This is supposed to be one of the drugs employed by the Obi wizards and witches. The most horrible rites, accompanied by atrocious cruelties, were performed amongst certain West African tribes and are continued amongst their descendants, the freed slaves of the West Indies and of the Southern United States.

Even to-day no white man is allowed to learn anything of the proceedings, but some form of devil-worship or Shamanism, accompanied by incantations and the use of poisonous drugs, still flourishes. Preparations of various sorts of Datura or Thorn-apple produce sometimes stupefaction, sometimes frantic, furious delirium, and sometimes death.

It is used in medicine as a narcotic and diuretic. Burton says that the Arabs smoke the leaves in pipes as a cure for influenza and asthma. It is sometimes used in Europe for neuralgia and even epilepsy. On the other hand, the priests of the ancient Peruvians used Datura to produce the ravings mistaken for inspiration, and it is supposed that the priests of Apollo at Delphi employed an allied species for the same purpose. In India, China, West Africa, and amongst the American blacks, it is still very commonly used.

A firm belief existed in the Middle Ages that every plant was a good remedy for something. There is a real basis in fact for this superstition, because every plant in the world has, so far as it can do so, to protect itself. The attacks of all sorts of grazing animals, from the mouse to the elephant, as well as the infinitely more dangerous and destructive insects, bacteria, and fungi, have to be provided for. By far the commonest form of protection is to develop within the plant strong medicinal or strongly smelling substances. These are far better as protective agents than the thorns and spines characteristic of deserts and half-deserts. We have already glanced at the turpentines and resins of Coniferous forests and at the odorous gums, frankincense, and myrrh of the Acacia scrub.

The use of poisons as protection is eminently characteristic of three of the natural orders. The Buttercups (*Ranunculaceæ*), the Potato order (*Solanaceæ*) and the Lilies. Of the first named, the celery-leaved, and indeed all Buttercups, are extremely poisonous; so also are all Aconites and Hellebores, as well as Marsh Marigold, Adonis, Clematis, and Larkspur.

Others, though not poisonous, are strongly medicinal, such as Blake Snakeroot, Hydrastis, etc. It is therefore inadvisable to use any of this order for food unless other people have eaten it without any inconvenience!

The beauty of the Lily order does not prevent it from being a particularly dangerous group of plants. Perhaps the worst poisons in this order are those of the Meadow Saffron (*Colchicum autumnale*), Herb Paris, Veratrum, Sabadilla, Lily of the Valley, Tulip, and Crown Imperial bulbs.

Chamælirium, Trillium, Squills, Garlic, Solomon's Seal, Aloes, and the Sarsaparillas are all well-known medicines.

The order *Solanaceæ* is perhaps the most interesting, for it includes such dangerous poisons as Tobacco, Datura, *Atropa belladonna* (Deadly Nightshade), Henbane, Bittersweet (*Solanum dulcamara*), Common Nightshade (*Solanum nigrum*), and a very great many important drugs. Even the common potato contains a poisonous secretion *solanin*, and it is dangerous to eat green potatoes or the foliage. Yet the Tomato or Love Apple (so called because it was supposed to excite tender feelings) is both nutritious and delicious. Chillies and Cayenne Pepper (*Capsicum spp.*) are also commonly used as condiments.

Such poisonous orders should of course be avoided, but much more dangerous are those deadly plants which appear as it were accidentally in orders which are amongst the most useful friends of man. Amongst the grasses there is the deadly Darnel (*Lolium temulentum*), a first cousin and not very unlike the very commonest and one of the most useful grasses—Rye Grass (*Lolium perenne*).

Then in the useful Carrot order, there are such dangerous and even deadly plants as Fool's Parsley, Water Dropwort, and Cowbane. *Œnanthe crocata* (Water Dropwort) is one of the very commonest marsh and ditch plants in Great Britain. It is perfectly well known to botanists as distinctly poisonous, yet in 1902 a veterinary surgeon brought me some of the tuberous roots to name, and told me that six fine young cows were lying dead on a neighbouring farm through having eaten them!

A particularly useful order of plants (*Leguminosæ*), the Beans and Peas, contains a few poisonous species. It is said that in every year children are sure to be killed by eating the seeds of the Laburnum, and to this order belong also the Calabar Bean and Crab's Eyes. The last named is only fatal when introduced below the skin in small quantities. The seeds of the Bitter Vetch (*Lathyrus sativus*) produce paralysis of the legs in man and also in horses. The Crazy or Loco weed of North America is sometimes eaten by horses in the Western United States. The wretched animals stagger about as if intoxicated, and eventually die. Belonging to this same order is the Wild Tamarind, or Jumbai, of Jamaica (*Leucæna glauca*). It is a weedy-looking acacia, and extremely common in all tropical countries. Dr. D. Morris thus alludes to it:—

"Mr. Robert Russell, of St. Ann's, informs me that horses feeding on the leaves of this plant completely lose the hair from their manes and tails. This ... statement was supported by the testimony of so many people acquainted with the facts that there was no reason to doubt it. Many years afterwards (in December, 1895), I renewed my acquaintance with the plant in the Bahamas. The plant was much more plentiful there than in Jamaica; it was, in fact, distinctly encouraged in the former islands as a fodder plant. The people were fully aware of the singular effect it produced on horses, and added that it also affected mules and donkeys. Its effect on pigs was still more marked. These animals assumed a completely naked condition, and appeared without a single hair on their body. Horses badly affected by Jumbai were occasionally seen in the streets of Nassau, where they were known as 'cigar-tails.' Such depilated animals, although apparently healthy, were considerably depreciated in value. They were said to recover when fed exclusively on corn and grass. The new hair was, however, of a different colour and texture, 'so the animals were never quite the same.' One animal was cited as having lost its hoofs as well, and in consequence it had to be kept in slings until they grew again and hardened. The effects of the Jumbai on horses, mules, donkeys, and pigs were regarded as accidental—due to neglect or ignorance. The plant was really encouraged to supply food for cattle, sheep, and goats. The latter greedily devoured it and were not perceptibly affected by it. It will be noticed that the animals affected were non-ruminants, while those not affected were ruminants. The probable explanation is that the ruminants, by thoroughly mixing the food with saliva and slowly digesting it, were enabled to neutralize the action of the poison and escape injury. The seeds probably contain the deleterious principle in a greater degree than any other part of the plant. It was a common experience that animals introduced from other localities suffered more than the native animals. The latter were either immune or had learnt to avoid the plant as noxious to them."

That animals resident in a district are not poisoned by plants which are often fatal to sheep and cattle when on the march through it, has been often observed in Australia. The great "mobs" or droves of sheep passing slowly on their travels through the bush to a new district are often poisoned by the Caustic Creeper (*Euphorbia Drummondi*). "The head swells to an enormous extent, becoming so heavy that the animal cannot support it, and drags it along the ground"; but this does not apparently happen to resident cattle. Similarly for the Darling Pea or Indigo (*Swainsonia galegifolia*). At one place this was growing abundantly where some travelling horses were hobbled for the night. "They had been on the road some nine weeks, and were up to this date caught without any difficulty. On this occasion ... their eyes were staring out of their heads, and they were prancing against trees and shrubs.... When driven they would suddenly stop, turn round and round, and keep throwing their heads up as if they had been hit under the jaw.... Two out of nine died, and five others had to be left at the camp."

In other natural orders we find one or two dangerous plants amongst a whole series of perfectly harmless or useful forms. The Oleander, in the Olive order, Corncockle (*Lychnis floscuculli*), in the Pink order, *Lactuca Scariola* amongst *Compositæ* and others are all cases in point. So also is the Yew amongst *Coniferæ*, etc.

How do animals recognize these particular plants as being dangerous whilst all their allies are harmless? But the reader will answer that they do not; it is well known that animals *are* killed by eating poisonous plants, therefore poison cannot possibly be any protection against animals.

This is one of those interesting questions in which the suppression of apparently irrelevant details produces confusion.

As a matter of fact, wild animals, or even domesticated animals in nearly a wild state, do *not* eat the poisonous plants of the country in which they and their forefathers have been brought up—that is provided that they are either adult or are accompanied by full-grown animals. Almost every case of cattle-poisoning in Great Britain occurs when young calves, foals, or lambs are turned loose in the fields without any mature older head amongst them. Sometimes valuable stable-bred animals are lost, especially by eating yew-leaves, but there are exceedingly few instances of full-grown cattle being caught in such foolishness. When cattle, horses, or sheep are turned loose in a new country, plenty of cases do occur, and it is possible that they might make mistakes with unknown foreign plants which had escaped into their pastures here.

But almost every case of poisoning, even of cattle, shows that it is young cattle who foolishly eat foxgloves, dropwort, buttercup, etc., and occasionally die thereby.

Wild animals, who are of course brought up by their mothers, never seem to be poisoned. They probably recognize the dangerous plant by colour, smell, or taste. As a matter of fact, many are rendered conspicuous by some lurid sort of colour, such as bright red or purple. There is a general garishness of appearance about many of them. Aconite, Foxglove, Herb Paris, Henbane, and Nightshades all show this peculiar appearance. In Java it is said that the natives keep away wild pigs by planting hedges of certain species with purplish-red leaves around their plantations.

Perhaps the most interesting point of all is that it seems to be quite justifiable to conclude that animals do, somehow, manage to tell their offspring and each other what they should and should not eat.

Youth, with its tendency to rash experiment, is thus kept in check by the mature experience of age.

But it must be admitted that it is exceedingly difficult to arrive at the facts in any particular case.

I shall be rash enough to give an opinion as to the actual facts in connexion with the common Yew (*Taxus baccata*). The seeds are poisonous to poultry and pheasants, but the fleshy part round the seed is eaten with impunity by many wild birds (blackbirds, etc.). The leaves are sometimes poisonous and even fatal to horses, cattle, sheep, donkeys, and goats, but they are not eaten by or are harmless to roedeer. When, however, e.g., horses are killed by eating yew, it is generally found that they have been grazing on cut-off branches which have been left lying on the ground. In this condition probably some specially poisonous substance is developed in them.

As regards rabbits, it would be extremely comforting to believe that they would eat yew-leaves or anything else which would kill them, but, so far as one can judge, they can eat all sorts of things which ought to do so with perfect impunity.

CHAPTER XIX. ON FRUITS.

BRIGHT COLOURS OF FRUITS—UNRIPE FRUITS AND THEIR EFFECTS—AN INTEMPERATE FUNGUS—ORANGES—PRICKLY PEAR AND THE MONKEY—STRONG SEEDS—BILL-OF-FARE OF CERTAIN BIRDS—A WOOD-PIGEON AND BEANS—ANTS AND SEEDS—BATS, RATS, BEARS, AND BABOONS—THE RISE IN WEIGHT OF A BIG GOOSEBERRY—MR. GIDEON AND THE WEALTHY APPLE—CROSSING FRUITS—BREADFRUIT AND BANANA—DATES—FIGS—OLIVES—PINEAPPLES BY THE ACRE—APPLES AND PEARS—HOME AND CANADIAN ORCHARDS.

AT Christmas time and during late autumn, there is but little colour in the country. Most green grasses have become a dull greyish-green, and the leafless brown and grey branches of the trees are not, at first sight, particularly interesting.

But amongst this monotony of sober colouring, points of bright red or flaming scarlet may be noticed here and there. Sometimes it is a spray of Hips (the fruit of the Rose), or it may be a cluster of Hawthorn berries. At Christmas the Holly is positively gaudy with its bright scarlet fruit set off by the shining dark green leaves.

Most fruits are some shade of red, but every fruit is conspicuous and easily seen.

This is one of the important harvests in some parts of Queensland.

There is the most extraordinary range in colour. The Snowberry and Dwarf Cornel are pure white. The Mistletoe is a yellowish green. Pure yellow fruits are not common, but some of the Cucumber orders and Lemons are lemon or orange-yellow. The bluish-black of the Blaeberry or Bilberry, of the Bramble, and of many Plums and Prunes, goes along with a rather peculiar shade of green in the leaves which sets them off. The black Elder berries, on the other hand, have bright red or pink stalks which contrast prettily with them. The colours of apples vary: many of them have been rendered a gorgeous, glossy red through cultivation. One of the most beautiful colour contrasts in Nature is found in the rich black of the Olive, with its background of shining white twigs and silver-green leaves. Another very curious harmony is that of the Spindle tree fruit, which has a hard dull red case that opens to display the seeds: these are enclosed in a bright orange fleshy cup.

Changes often occur. The Lily of the Valley fruit is at first green, then becomes flecked with red, and finally is a rich scarlet. Juniper berries change from green to purple.

Now there is always some meaning in Nature for any series of facts such as these. Why are these fruits so brightly coloured and so conspicuous?

Birds and other animals are intended to scatter the fruits and seeds, and so the fruits must be easily distinguished at a distance. The seeds are taken to some other place, where they germinate and form a new plant. This furnishes the clue and guide to many other peculiarities in fruits and seeds.

The pleasant smell of ripe apples, plums, strawberries, and other fruits, also attracts birds and other animals. But the sugary juice and delicious flesh is developed entirely for the purpose of making it worth a bird's while to eat it. The amount of sugary matter is enormous, and the seeds seem very small and inconspicuous compared with this luscious mass. The sugar is produced very rapidly towards the end of the ripening period.

A Cucurbita fruit, for instance, may increase in weight at the rate of ·0032 ounce per minute. All who have gathered strawberries know how quickly they ripen.

The way in which the sugar is formed is not understood, but unripe fruits contain bitter, unwholesome acids and essences which may produce colic or very unpleasant effects if the fruits are eaten green. Thus the colour is a guide to the animal, who is not supposed to eat the fruit until it is ripe; if eaten green, the seeds inside the fruit are quite destroyed and cannot germinate. Yet animals are so greedy that young birds, young animals of all sorts (even girls and boys) will and do eat green or half-ripe fruit. In this present year there is no doubt that many children have suffered for having done this. Yet if we come to think of it, throughout all the millions of years during which fruits have ripened, Nature has every year clearly told young pterodactyls and other lizards, young birds, young monkeys, and young people to wait till the fruit is ripe. None of them have learnt to do so.

When investigating by experiment, on the vile body, the properties of plums, strawberries, and other fruits, you are sure to find here and there one that has decayed and become rotten. In most cases this is because a bird has pecked a hole in it, or because the outside skin has been broken by a wasp. The sugar has then begun to ferment. Why does it do so?

If you gather a few fruits, put them into a jar of sugar-water, and leave it after closing the mouth with a bunch of cotton wool, then in a day or two fermentation begins and alcohol is produced. That is because, on the outside of the fruit, there were hundreds of an objectionable little fungus. It lives upon sugar and turns the latter into alcohol. This yeast fungus is really a living distillery. It lives in the midst of alcohol all its life, dying eventually (like the Duke of Clarence in his butt of Malmsey wine) by alcoholic poisoning, which it has brought about by its own work. This little yeast fungus can only be seen with a microscope. From a rotten fruit it drops on to the ground, where it remains all winter. Next spring certain small insects (green-fly and the like) carry some of these yeasts from the earth to next year's fruits. But the skin of the plum or apple, or the hairs on a gooseberry, or the delicate, waxy bloom on a grape, will prevent these insects or wasps from laying open the sugar inside the fruit to the attacks of yeasts and other fermenting fungi.

Some fruits appear to have "favourites"; they seem to prefer that large animals should eat them. If you look carefully at a piece of orange peel, and cut a small piece across, you will see distinctly small resin pits full of a curious essence which gives the characteristic taste to marmalade. This bitter stuff will prevent wasps from touching the sugar. It is, however, a valuable material, and some kinds of lemons, etc., are grown chiefly for this oil, which is obtained by scraping the peel with a little saucer which is studded with short pins.

A still more extraordinary fruit is the prickly pear; this is very delicious though very difficult to eat. Indeed, only monkeys and man seem able to enjoy it. The sugary part and the seeds form a little round mass in the inside. The outside part, though also fleshy, contains hundreds of minute mineral needles, which stick in the tongue and lips and cause most painful inflammation. The monkey eats the prickly pear with very great caution, getting his fingers into the top and scooping out the sugary part. Man requires a teaspoon to do this satisfactorily.

Another very curious point about these fleshy fruits (and also ordinary ones) is the strength of the seed inside. It does not look very strong.

But an orange seed, for instance, will not be in the least injured if you put it between two glass plates and gradually press upon the upper one up to even a pressure of some thirty pounds. Even hemp seed, which seems quite weak, will endure a weight of four pounds. It is impossible to break a prune stone, or to injure a date stone, by standing with your whole weight upon it.

Such strength is necessary because many of these seeds are eaten by birds and ground up in their crops with bits of china, stones, shells, and the like, which the birds pick up just to help them in crushing their food.

Fruits and seeds would seem to be exposed to some danger when they are lying on the ground. Horses or other heavy animals might tread on them. But the strength of seeds and their shape is such that no harm is likely to accrue. For instance, I arranged a thin layer of garden earth (a quarter of an inch thick) on a glass plate; upon the earth I placed four hemp seeds; then I put a 58-lb. weight on the top of the seeds. They were not in the least injured, although the seed of the hemp is not a particularly tough one. Under such conditions the seed simply slips into the earth.

This is made easy for it on account of its shape, which is generally rounded above and below. A transverse section of a seed would be in shape like the arch of a bridge and its shadow in the water, at least in many cases. There are also usually wonderfully thickened cells in the shell or coat of a seed, which makes it tough and strong.

The following are a few cases of strong seeds or fruits: —Cotton seed bears a weight of 19 to 20 lb.; the hard fruits of the Dogrose, 33 lb.; Castor-oil seed, 17 lb.; Hornbeam nuts, 27 lb.; Pine seed (various sorts), from 11 to 2 lb.; Yew seeds, 16 lb.; Peas, 50 to 56 lb. In every case they are not at all hurt by these pressures.

As regards the animals for whom fruit or seeds are of great importance, birds are of course the commonest. The following is part of the bill-of-fare of a few of our common birds:—Thrushes eat blaeberries (bilberries), brambles and mulberries. Missel-thrush (or mavis) is especially fond of the mistletoe.

Now the berry of the mistletoe is exceedingly sticky and glutinous, and in the course of the bird's meal these sticky strings get on to the bill and feathers, so that the mavis wipes its bill on the branch of a tree. When it does so the seed becomes attached to the branch, and is drawn close to the latter when the viscous matter dries up, and so takes root on the branch.

Nightingales and robins eat strawberries and elderberries; blackbirds are very fond of strawberries, gooseberries, and raspberries. Wood-pigeons eat beechmast, acorns, and, according to Pliny, mistletoe-berries also, but this latter author has not been confirmed by later observers. Some of the wild African pigeons are exceedingly fond of castor-oil seeds. When travelling through the Central African bush, it is often necessary to shoot your dinner (if you are to have any at all), and castor-oil bushes can be relied upon to produce pigeons, if you are content with and are able to shoot them.

There is a widely-spread belief in the country that a great quantity of berries means that a very severe winter is going to follow. But as a matter of fact the winter of 1904 was not a severe one, and yet there were enormous quantities of berries.

We are still ignorant of many details about birds and berries. It is not quite clear how the seeds are not destroyed, though experiments have shown that they are not injured, by passing through the body of a bird. Kerner von Marilaun, for instance, tried the fruits and seeds of different plants which were offered to seventeen birds, as well as to marmots, horses, cattle, and pigs. He found that from seventy-five to eighty-eight per cent. of the seeds germinated afterwards so far as regards the blackbird, song-thrush, rock-thrush, and robin. Quail also bring seeds from Greece and the Ionian Islands to Sicily.

Mr. Clement Reid says: "Some years ago I found ... in an old chalk-pit the remains of a wood-pigeon which had met with some accident. Its crop was full of broad-beans, all of which were growing well, though under ordinary circumstances they would have been digested and destroyed." Such accidents are common.

But it is not only birds which eat fleshy fruits and seeds. Even the tiny, industrious ant drags about seeds of certain plants. Sometimes they gather up corn or grasses, such as ant-rice, and store them for use in winter. They even bite off the growing root to prevent the seeds germinating and spoiling. Occasionally they seem to carry the seeds by accident, as, for example, those of the cow-wheat and a few others which resemble their cocoons in size, colour, and form. In other cases there is a little fleshy excrescence on the seed which they are fond of eating. Cyclamen, snowdrop, violet, and periwinkle seeds are supposed to be carried in this way. Many animals occasionally or regularly eat fruits. There are, for instance, the flying-foxes or fruit-eating bats of Madagascar and tropical countries, which may be seen hanging from the upper branches of trees by their toes, with their heads tucked away under their wings. When disturbed a little fox-like head appears, and after much chattering, scolding, and expostulation, the creature unhooks itself and flies away with a strong flight not unlike that of a crow. Horses are occasionally fed on peaches in Chile. Rats eat the coffee cherry, and do a great deal of harm in coffee plantations.

In Cashmir the mulberry and other fruit trees are sometimes visited by sportsmen, who often find bears feeding on the fruits. Pigs, of course, eat all sorts of fruit, and several other mammals do the same, but it is especially monkeys that live chiefly on fruit. They plunder the banana plantations, and in South Africa melon-patches require to be most carefully watched to prevent baboons from destroying them.

It is said that the baboons watch the plantations from a distance, and will only come down if they think no one is there: so five people walk to the patch, and while four go away again, one of them remains in hiding to shoot the baboons, who cannot tell the difference between four and five.

Man himself is, and has always been, a great eater of fruit. Not only so, but he has enormously improved and altered wild fruits until they are modified into monsters of the most extraordinary kind. The ordinary wild gooseberry weighs about 5 dwt. But even in the year 1786 some of the cultivated forms weighed double this amount (10 dwt.), and in 1852 gooseberries which weighed more than 37 dwt. were in existence. What size the largest big gooseberry may be this year is not very easy to say, because the public Press is at slack times too energetic about the question. The most usual way of improving fruits is by selecting the finest specimens for reproduction. It is by this means that the original wild banana, which is a rather small fruit with very large seeds and very little flesh, has been altered into something like 150 varieties, of which the immense majority have no seed at all. This is a very extraordinary fact, because the seed is the reason for the existence of the fruit. Of course, all such varieties must be reproduced by suckers (like the banana) or by grafts, or in some such non-sexual manner. Seedless varieties exist of the Cucumber, Fig, German Medlar, Diospyros, and Orange.

In the case of seedless varieties of the Vine, it has been found that it is necessary to carry pollen to the flowers to fertilize them, and the seedless fruit is also very much smaller in this case, not more than a quarter of the size of one that has seeds.

The following instance is typical of the manner in which many well-known kinds of fruit have been developed, though the perseverance shown by Mr. Gideon is certainly not common. About the year 1855 this gentleman began planting apple trees of about thirty named varieties. For nine years he continued his experiments. He not only planted trees, but also sowed apple seed sufficient to produce a thousand trees every year. Yet the cold winters were so severe that at the end of ten years one small seedling crab apple was the solitary survivor. One seedling of this turned out to be hardy enough for the climate of Minnesota, and this, the "wealthy" apple, has been of great importance to the Northern Mississippi growers. It is to be hoped that the name has been justified in Mr. Gideon's case.

A West Indian negro thinks nothing of walking twenty miles with loads such as these.

Many other cases could be mentioned of a chance variety produced as a wild plant, and then propagated non-sexually for long periods, e.g. the New Rochelle Bramble, which was found by the roadside, and which turned out to be exceedingly valuable. It is by crossing or hybridizing that the most extraordinary results have been obtained. Sometimes with plums, the hybrids of the first generation are nearly double the size of their parents. Some of the crosses are between different plants. The Loganberry, for instance, is said to be a cross between a Raspberry and a Bramble. It ripens in July, and is said to be far in advance of either of its parents as regards juiciness and acidity.

In most cases, however, the crosses are between well-established varieties or races of the same species, and both hybridizing and selection are employed to get the desired result.

There are several tropical fruits which, with the possible exception of wheat and oats, are more important to mankind than anything else. The Breadfruit (*Artocarpus incisus*), which is very common in the South Sea Islands, has a large fruit the size of a melon. When baked in an oven heated by hot stones, it forms a satisfying meal: it is rather like new bread, but has very little flavour. Coarse cloth is made of its bark, and the wood is used as timber. The tree also has a milky juice containing indiarubber, and is employed for caulking the canoes. The most interesting point for botanists about this plant is that the fruit is made up of thousands of little flowers, and the fleshy part is really the stalk. Fossil trees of this genus (of the chalk period) are found in some parts of Europe.

Still more important to mankind is the Banana (*Musa paradisiaca*). It is wheat, corn, and potatoes all in one, in tropical and sub-tropical countries. It is found all over the world wherever there is a hot, moist climate and shelter from wind. It is a most generous plant as regards the amount which it will produce. It will yield about 19-1/2 tons of dry fruit on a single acre, which is about forty-four times the amount given by potatoes and 133 times that of wheat. Moreover, it differs from almost every other fruit in being both "rice and prunes," that is, it is nutritious and wholesome, and yet at the same time succulent. There are still people who declare that the taste is that of "cotton wool and Windsor soap," but that is a frivolous and unjust remark. It is very difficult to prepare it exactly in the right way for export to Great Britain, and the slightest change in temperature or period of gathering has the most distressing results.

As with many other tropical fruits, the countries where it is most carefully produced and where the trade is most important are just on the borders of the tropics. There Europeans can keep enough vigour and vitality to supervise and watch over the labour of natives. It is in the Canary Islands, Queensland, and Jamaica that the cultivation is most carefully looked after. The yield may be from five hundred to a thousand bunches per acre, and the value of the trade is enormous. A plantation is not very beautiful, because the huge leaves break up into irregular, ragged pieces which look untidy. The flowers are visited by the beautiful little honey-sucking sunbirds and humming-birds. Monkeys also are very fond of the fruit.

In the tropics it grows everywhere, and with extremely little trouble. It is a doubtful blessing to the negroes, for they get their food so easily that they tend to become incorrigibly lazy. Jam, champagne, brandy, and meal can be made from the banana. When this meal can be prepared satisfactorily, it may partly replace wheat in temperate countries. Besides this, the leaves are used for thatching, and the stalks which make the stem contain a valuable fibre which is used for string and rope.

In Egypt and all along the great deserts of Sahara and Asia the graceful stately Date palm gives the favourite food of the people (see Chap. X.).

The Arabs grind up the stones to make food for camels, and sometimes ferment the sap to make toddy. The trees are either male or female. The Arabs knew that it was necessary to pollinate the female flowers with male pollen long before the meaning of the process was realized in Europe.

The Fig, a native of the Persian Gulf, is cultivated all along the Mediterranean and in India, Australia, and California. It is sometimes fifteen to thirty feet high, and reaches a very great age. There is one at Finisterre said to be several centuries old. It yields fruit worth about £14 an acre. The most interesting point about the Fig is the way in which the Fig-wasp carries the pollen (see Chap. V.).

Olives are also one of the most important and characteristic Mediterranean trees. The crop in both Spain and Italy is worth about £8,000,000 to £9,000,000 annually. In California it is also successfully cultivated, and pays very well. The peculiar taste of the dessert olive is obtained by soaking it in lime or potash, and then in vinegar or salt.

The Pineapple is one of the most delicious fruits, and is interesting in every way. The little sharp spines on the edges of the leaves keep animals off, and also make it a little difficult to harvest. The workmen must wear leather trousers to prevent their being cut and torn by the leaves. In Queensland the pineapple is grown in big fields, and about ten thousand fruits (worth about one penny each) can be got from a single acre. It is also grown in the West Indies, in India, and in other tropical countries. If you examine the horny outside skin of the fruit with a sharp penknife, you will find that each little piece of the mosaic is a flower in itself; with a little care the bracts, three sepals, three petals, and six stamens can be distinguished. The whole stem and all its flowers unite to make a compound fruit. Most varieties have no seeds. It is a native of South America.

It is, however, our home fruits, Apples, Pears, Gooseberries, Strawberries, Raspberries, and Currants, that are most important to us in Britain. The Wild Crab Apple is found from Drontheim, in Norway, to the Caucasus, and grows over the whole of Europe. Apples were known to the Greeks and Romans.

Unfortunately, in our own climate there are great dangers in the orchard. A touch of frost when the flowers are ripe will very likely kill the tender, green, baby apple. It is perhaps in Canada and North America that the growing of apples and pears is most carefully looked after. Our beautiful old orchards in Devonshire and other places, with comfortable grass below the trees, and moss-covered, picturesque, ancient trunks, are not found in the New World. The regular lines of young trees in bare, carefully-kept earth, with every stem whitewashed and treated with the most scientific monotony, produce a most valuable return. But in this country those who are careful and scientific sometimes obtain extraordinary results. It is on record that a man with a holding of twenty-nine acres near Birmingham made £600 a year from this small plot and paid £250 for labour on it.

Mr. Gladstone also said that the future of British farmers depended upon jam. Yet it must be remembered that the trees take a long time to come into bearing, and the crop is most uncertain.

CHAPTER XX. WANDERING FRUITS AND SEEDS.

SHIPS AND STOWAWAY SEEDS—TIDAL DRIFT—SHEEP, BROOM, MIGRATING BIRDS—CROWS AND ACORNS—ICE—SQUIRRELS—LONG FLIGHT OF BIRDS—SEEDS IN MUD—MARTYNIA AND LIONS—THE WANDERINGS OF XANTHIUM—COCOANUT AND SOUTH SEA ISLANDS—SEDGES AND FLOODS—LICHENS OF ARCTIC AND ANTARCTIC—MANNA OF BIBLE—THE TUMBLE WEEDS OF AMERICA—CATAPULT AND SLING FRUITS—COW PARSNIPS—PARACHUTES, SHUTTLECOCKS, AND KITES—COTTON—THE USE OF HAIRS AND WINGS—MONKEY'S DINNER-BELL—SHEEP-KILLING GRASSES.

THE ways in which fruits and seeds are scattered abroad over the face of the earth form one of the most fascinating chapters in the story of Plant Life.

There is an infinite number of ingenious contrivances, so many indeed that it is not at all easy to explain them.

However, suppose yourself seated on a grassy cliff near Eastbourne or Brighton.

Looking lazily out over the blue waters, you see Norwegian timber ships and steamers of all kinds, from the little coasting "Puffing Billy" to the huge liner departing for Australia or South Africa.

Plants are probably using every steamer; in the straw of the packing cases, in the cargoes of corn or grain, in the ore, and in the ballast, there are sure to be seeds. Such stowaways are mostly weeds, but of course many valuable garden, farm, orchard, and forest seeds are being intentionally exported.

Looking down on the seashore, you will notice the high-water mark, a yellowish brown line of floated rubbish which is quite distinct even at a distance. If you now go down and examine it closely (not a particularly pleasant operation, seeing that so much is in a decomposing condition) you will find many seeds amongst the corks and bits of straw, seaweed, and objectionable, if lively, animalcula, and very likely also pieces of plants, such as willow branches, which might quite easily take root.

On the coast of Norway, and on our own western seaboard, the fruits of a West Indian bean (*Entada scandens*) are occasionally to be found, and its seeds are probably able to germinate. We know that in long-past geological ages they were floating round the estuary of the Thames, where they occur as fossils. It has been found by experiment that fruits and seeds are not killed although they have floated for a year or more in salt water. Thus ocean currents are utilized to carry fruits and seeds.

But from our comfortable seat on the South Downs, still more can be learnt of wandering seeds. The wind which blows across the downs carries with it hundreds of winged or hairy fruits, all of them exquisitely fashioned as miniature airships, aeroplanes, or other winged contrivances. The wind is an important distributer of seeds.

One of the South Down sheep is trailing behind it a piece of bramble which has caught in its wool; others, which have been grazing on the broken cliff-edge where Agrimony, Forget-me-not, and Burdock are flourishing, are certain to have spiny or sticky fruits entangled in their wool. Animals therefore carry seeds in their wool or fur. If it should happen to be a fine, sunny afternoon, and if there are any plants of Broom near by, it is quite likely that you may, every now and then, hear a faint, sudden crack. This will be the Broom at work scattering its seeds by itself. The little pod, when it dries, contracts in such a way that it splits with a sudden explosive pop, and the seeds are sent flying to a distance of three or four feet. This curious fact was observed in 1546 by the naturalist Boek. The Whin and many other plants act in the same way, for the dry fruit becomes elastic and coils up spirally, flinging away the seed.

But here also, on the southern shore of England, we are at a main station of arrival and departure for migrating birds. A Landrail or other marsh bird might be flushed in France, and might quite easily cross the Channel with French mud sticking to its plumage. In this mud, or in its crop, there may be seeds or fruits which will be left in an English pond. This method is probably a very important one, for these plants growing in duck-haunted places are amongst the most widely distributed of all.

Mr. Reid has a very interesting discussion on this point. The crow or rook could quite well cross the British Channel now. In the days when Britain was covered with ice and snow, the gap between the French and the English shore was only half the present width. There was at that time much flat land with oak forest bordering the French coast.

Mr. Reid shows that it is probable that rooks regularly carry about acorns in the cup, for he found seedling oaks associated with empty acorn husks, stabbed and torn in a peculiar way. "On October 29th of 1895, in the middle of an extensive field, bordered by an oak copse and scattered trees, I saw a flock of rooks feeding and passing singly backwards and forwards to the oaks. On driving the birds away, and walking to the middle of the field I found hundreds of empty acorn husks and a number of half-eaten, pecked acorns." So that crows may have brought the acorns that colonized Britain with oak forest in the earliest historical period.

Another means of dispersal is not so obvious on the South Downs. In the Arctic region a glacier breaks away at its tongue into icebergs, which float off and are stranded somewhere perhaps hundreds of miles distant. Upon these icebergs are stones and soil and plants which may be carried to a great distance from their original place. In the Glacial period or Great Ice Age, ice may have been an important help in distributing plants, but at present it is difficult to find a good example.

From all this it is clear that in order to carry plants to new countries and new homes, everything that moves on the earth's surface can be employed. Not only the wind, but ocean currents, river waters, icebergs, and floating ice are used. Migrating birds, mammals, and especially the most restless and unsettled animal of all, viz. man, are at work consciously and deliberately, or unconsciously and accidentally, carrying the seeds to form new forest, grasslands, or harvests in other countries.

The subject is in truth so vast that it is difficult to select the most interesting and important cases.

The way in which squirrels, rats, voles, and lemmings devour nuts and the like often leads to the distribution of the fruit. A squirrel may, like a human being, forget where its store was buried, or be driven from the place. Then some of those forgotten nuts will grow into trees.

Birds are known to travel enormous distances. It is said that one little Arctic bird travels from Heligoland to Morocco in a single flight. It would not, at first sight, seem likely that seeds and fruits could be carried by birds; yet Darwin saw that this might possibly be the case. The mud and slime in which so many birds find the small insects which they require is full of seeds. An Austrian botanist, Kerner von Marilaun, examined the mud scraped from the beaks, feathers, and legs of a number of wading and marsh-birds. He found in it the seeds of no less than thirty-one different water and marsh plants (Grasses, Sedges, Toad-rush, etc.). This showed, as is very often the case, that Darwin was the first to discover a very important point. It is also interesting to find that these ugly little freshwater mud and marsh plants are at home almost everywhere, from the Arctic circle to Tierra del Fuego and from Peru to Japan.

The most extraordinary cases known of sticking fruits and spines are the Martynias and Harpagophytons of South Africa. The fruit is covered by hooked claws, and becomes a regular pest wherever it occurs. Deer, antelopes, and other animals get their hoofs entangled in the fruit, and the wretched creatures have to limp about until the hard thorny fruit is trodden to pieces. Dr. Livingstone says that the fruit gets into the nostrils of grazing animals which cannot possibly remove it themselves, and so have to wait patiently till the herdsman comes to take it out. According to Lord Avebury, lions may sometimes be destroyed by these horrible fruits. When a lion is rolling on the sand, the claws (an inch long) stick in his skin, and when the lion tries to tear it away with his teeth his mouth gets full of the fruits and he cannot eat, and perishes miserably of starvation.

Some of our common British fruits are most perfectly planned to stick or entangle themselves in the wool of sheep or in people's clothes. These, such as the Goosegrass (Robin-run-the-Hedge), Burdock, Forget-me-not, Sanicle, Avens, etc., have very often been described. It is only necessary to examine one's clothes after a walk through rough, broken ground to discover some of them, and the ingenuity and neatness of their tiny hooks, harpoons, or prongs can then be realized. We shall give one or two instances of some other spiny plants. There is, for instance, Xanthium, which is one of the Daisy flowers or Composites. Unlike most of this order, its little fruits possess no wind-hairs. The outside of the head of flowers is covered by strong curved little crooks. These get so entangled in wool or hair that they become a perfect pest to wool merchants. In 1814 Xanthium was unknown in the Crimea, but by 1856 it had covered the whole of the peninsula. In 1828 the Russian cavalry horses brought it on their manes and tails into Wallachia, from whence it travelled to Servia. Servian pigs carried it into Hungary. In 1830 it was taken in wool to Vienna. By 1871 it had reached Paris and Edinburgh. In 1860 Frauenfeld saw horses in Chile whose manes and tails were so felted together with thousands of these fruits that the animals could scarcely walk. In Australia, where it first appeared in 1850, it has caused a very serious loss to the wool merchants and squatters. The loss has been put at 50 per cent. by some authorities.

We have already alluded to the transference of fruits and seeds by ocean currents. In the *Challenger* expedition, no less than ninety-seven kinds of marine floating fruits were observed.

Amongst these the most important is the Cocoanut. The nut sold in this country is not the whole fruit, but only the inside shell. In the natural state this is enclosed in a dense mass of fibres, which form the valuable "coir" used for brushmaking and a variety of purposes.

The entire outside of the fruit is covered by a smooth white skin. The whole fruit is about the size of a man's head, and is so light that it floats easily in the water. It has in fact been carried by the waves to uninhabited islands all over the South Seas. It is a very great blessing to Polynesia, for a tree yields thirty to fifty nuts, and four of these nuts will furnish enough food for one day. Coprah and the oil extracted by boiling the inside are also valuable. Spirit or toddy can be made from the young buds. The leaves are used for thatching and the trunk for timber.

There are other very curious palm fruits which are also carried by water. Sir Joseph Hooker mentions the large, round fruits of Nipa, as big as a cannon-ball, turned over by the paddles of the steamer in the muddy waters at the Ganges mouth (*Himalayan Journal*).

In this country a search in the rubbish left by a spate or freshet along a riverside is sure to furnish many floating fruits or seeds. Most of these are small and rather difficult to see. Perhaps the most interesting are those of the Sedges. The real fruit is only about one-sixteenth of an inch in size, but it is enclosed in a little sack or bag a quarter of an inch long and with a narrow opening, so that it floats quite easily. Many willow branches, pondweeds, hornweeds, and the like, are also found in the rubbish left by floods, and these can often take root.

It is, however, in the exquisite modifications of those fruits which are blown by the wind that we find the most beautiful contrivances of all. They are effective also. Seeds are often so small as to be like dust particles, and such may be carried in the air to almost incredible distances. That of *Goodyera repens* weighs only 1/200,000,000 of a pound, that of *Monotropa*, ·000,000,006 lb. It is no doubt by the wind that the spores of lichens are carried from one mountain to another. On a map of the world the distance from the Arctic to the Antarctic, between the North and South Poles, seems enormous. Moreover, the amount of water, desert, tropical forest, and cultivated land in this extent of country is very great. There are but few rocks on which lichens could manage to grow. And yet of the Antarctic Lichens in the South Polar regions, and which are also European species, more than 73 per cent. are found in the Arctic or North Polar regions.

An Arctic lichen spore probably travelled from Scandinavia to the German and Swiss Alps, another journey took it to the Atlas Mountains, thence to Abyssinia, again to Mount Kenia, and from there, somehow, it wandered to the South Orkneys or King Edward VII Land.

While talking of lichens, one must not forget the Manna of the Bible (*Lecanora esculenta*) and two other species, which form warted, wrinkled masses on rocks. It breaks off and may be carried away by the wind, or in heavy rain it may be washed into depressions of the soil, where a man can pick up 8 to 12 lb. in a day.

It "is used as a substitute for corn in years of famine—being ground in the same way and baked into bread.... It is also remarkable that all the great so-called rains of manna, of which news has come from the East to Europe, especially those of the years 1824, 1828, 1841, 1846, 1863, and 1864, occurred at the beginning of the year, between January and March, i.e. at the time of the heaviest rains.... The inhabitants of the district actually thought that the manna had fallen from heaven, and quite overlooked the fact that this vegetable structure grew and developed (although only in isolated patches and principally as crusts on stones) in the immediate neighbourhood of the spots where they collected it."

Amongst the wind-blown fruits and seeds there are cases in which entire plants are dragged out of the soil and hurried away by the wind, which rolls them over and over. They may be blown along for days together. The seeds drop out by the way. In this country one rarely sees anything of the sort, but in the Prairies of North America, when under cultivation, these tumble-weeds are a serious and expensive pest. Sometimes the farmers dig trenches to catch them, or they may put up fences against which the tumble-weeds become piled or heaped up until they blow over the top.

It is not very much use to give the names of these weeds, for they are mostly rare or not British species. Such tumble-weeds are generally nearly spherical in general form and have a short, rather weak, root which is easily torn out of the ground. In some grasses, such as "Old Witch," a well-known pest of the United States, the grass-stalk, with many flowers on it, is pulled out of its sheath and blown away.

But it is more usual for the fruits or seeds themselves to break off the parent plant, and to be carried away by the wind. To this end we find the most extraordinary changes. Although the flower may droop from its stalk, the latter becomes upright and grows quite a considerable length when the seed or fruit is dispatched on its wanderings. This will raise the fruit or seed as high as possible above the surrounding grasses.

Then in some cases the fruit opens to allow the seed to escape. Small holes appear in it, or the fruit splits. As the dry, elastic, withered stalk swings to and fro in the wind, the seeds are swung out of these openings, and starting with a certain momentum the wind will carry them often to a surprising distance from their parents. In wet or rainy weather these holes or slits generally close together, and no seeds are sent forth on their travels. The little holes in the top of a poppy-head by which the seeds are swung out have little flaps, which close over and shut them up in wet weather.

Some plants make a sort of catapult to sling or hurl their fruits. Kerner von Marilaun was the first to describe some of these curious arrangements. He had brought home some fruits of *Dorycnium herbaceum* and laid them on his writing-table. "Next day as I sat reading near the table, one of the seeds of the Dorycnium was suddenly jerked with great violence into my face." Some of the neatest catapult fruits are those of *Teucrium flavum*. (There is a British species, the Woodsage, but it has not got the same arrangement.) When the petals have fallen off, the four small fruits are left inside the cup-like sepals; the flower-stalk when dry is very elastic, and if an animal touches the sepals it swings violently and shoots out one of the fruits. But that is by no means the whole of the process: there are hairs arranged spirally in the throat of the sepals, and these give a spin or twirling motion like that of a rifle-bullet to the fruit. The fruit also flies out of the sepals in a line of flight which is inclined at an angle of about forty-five degrees to the horizon; at this angle, as is well known, the trajectory or distance travelled will be the greatest possible.

But by far the best way to understand these questions is to try with some common weeds in the country towards the end of summer or beginning of autumn. If either the Cow Parsnip or wild Angelica, or Myrrhis, be gathered and kept till it is quite dry, then if you take it by the stalk and swing it to the full extent of the arms the fruits fly off to fifteen (or more) feet away. Every part is elastic—not only the main stalk, but the thin separate stalks of the flowers and also the delicate piece by which each half-fruit is attached. The half-fruits themselves are also so made that they are of exactly the right shape to take a long flight.

Ever since the days of Icarus, one of the unsatisfied ambitions of mankind has been to fly like a bird, to "soar into the empyrean," and to be no longer chained to the earth's surface.

It is a very curious study, that of the many and diverse inventions, almost always useless and very often fatal, by which men have endeavoured to solve this problem. Every one of these can be paralleled amongst the many neat contrivances of wind-borne fruits and seeds. The principle of the "parachute," which is more or less like an umbrella, is found in both fruits and seeds. One of the most beautiful is the Dandelion fruit, where a series of the most exquisite branched hairs springs from the top of the slender shaft which carries the little hard fruit. Most of the Composite or Dandelion order have, however, more of the "shuttlecock" idea. There is a row or crown of stiff and spreading or feathery hairs.

Negroes picking the cotton harvest.

The classical person above alluded to seems to have copied the bird's wing, sticking on feathers with wax, which of course melted in the sun with the usual result to the inventor of flying machines. Many seeds have regular wings which act like those of the bat or flying squirrel. One of the most exquisite of all is the seed of Bignonia. The Dahlia fruit has also a flying wing, and a great many others might be mentioned. Major Baden-Powell experimented with kites, which were supposed to raise a man high enough in the air to take observations of the enemy's movements. But a most exquisite "kite" is that of the Lime tree. The little fruit is hung from a broad, flying bract, and as it very slowly sinks to the ground it solemnly turns round and round. That is because the pressure of the air acts on the flat bract just as it does on an aeroplane, and forces it to revolve. So the fruit remains a long time in the air, and may be carried to nearly a hundred yards away from its parent tree.

The Traveller's Joy (*Clematis*) and the Cotton have their seeds covered all over by many entangled hairs, which act like a piece of fluff, so that the wind blows the seed away.

No one has discovered the original wild Cotton plant. The robes of the priests in Egyptian temples were made of it. It was introduced into Spain by the Arabs when they invaded that country. When the Spaniards attacked the half-civilized Indian people of Central and South America, they found cotton was regularly cultivated there. Its history in England is rather interesting. In the days of Queen Elizabeth the great English industry was the production of woollen cloth from Yorkshire sheep. A penalty of £20 was imposed, even as late as 1720, on any person who imported or even wore cotton cloths. Yet this was unable to stop the growth of the trade which, thanks to the Flemings and Huguenots who took refuge from religious persecution in this country, eventually became our gigantic textile industry employing millions of factory hands.

The advantage of these wings and hairs is at once seen if one compares the time that a fruit or seed takes to fall through a given height, first with its wings or hairs, and then after they have been cut off.

An Artichoke fruit, for instance, will take nearly eight seconds to reach the ground from a height of a few feet. But if you cut away its hairs, it will touch the ground in a little more than one second. A Sycamore fruit of which the wing has been removed falls to the ground in about a quarter of the time that it takes when it has not been injured, so that the wing helps it to fly to four times the distance that it could reach if it had none. The Ash fruit also remains twice as long in the air as it would do if it had no wing; and so on.

We shall finish this chapter by describing two very extraordinary cases.

The Sandbox tree is a native of tropical America. The fruit, as large as an orange, consists of a number of rounded pieces, each with a single seed inside. When ripe each piece splits off, making a noise like the report of a pistol. The plant is sometimes called the Monkey's Dinner Bell. These pieces may be thrown to a distance of fifty-seven feet from the parent plant.

Even more remarkable are the hygroscopic grasses. There are four of them, which are widely separated as regards distribution, for one (*Stipa capillata*) lives in Russia, another (*Stipa spartea*) in North America, a third (*Aristida hygrometrica*) is found in Queensland (Australia), and the fourth (*Heteropogon contortus*) belongs to New Caledonia.

Yet all these four grasses are said to kill sheep, and do so in a manner that is almost identical. The mechanism is as follows.

The fruit is like that of most grasses, enclosed in a folded leaf, the bract (or glume), which in these particular cases is produced into a very long fine tapering hair or awn. This awn is sensitive to changes in the *moisture* of the air. It is strongly hygrometric: in wet weather it straightens itself, and it coils into corkscrew spirals in dry weather. The widened part of the base, which contains the grain, tapers into a sharp, very hard point; upon this there are, on the outside, many stiff hairs, which point backwards away from the sharp tip.

Now, suppose this fruit to fall on the ground, the awn or tail is sure to be entangled in neighbouring grasses or herbs, but the hard point will rest upon the ground. Every coil and twist made by the entangled awn or tail will push the point a little deeper into the earth, and the backward-pointing stiff hairs will prevent its being pulled out of the soil.

Therefore all these modified contrivances ensure that the seed will bury itself.

But supposing that one of these fruits falls upon a sheep's back. Then an exactly similar process will go on. The seed will be forced through the skin into the body of the sheep. In fact, if it should fall above any soft or vulnerable part of the animal, the sheep will very likely be killed.

As a matter of fact, sheep are said to be killed by these grasses in all those four countries, distant though they are from one another.

We have endeavoured in this chapter to give some faint notion of the hundreds and thousands of ingenious contrivances utilized by plants in order to ensure the dispersal and future prosperity of their children.

Every species is always trying to colonize new ground, to seek fresh fields and new pastures. Plants are not content to keep to the old habitats, but every species tries to scatter its pioneers over all the neighbouring country, so that, as often happens, if it is exterminated or suppressed in one locality, new generations luxuriate elsewhere.

CHAPTER XXI. STORY OF THE CROPS.

BLOATED AND UNHEALTHY PLANTS—OATS OF THE BORDERERS, NORSEMEN, AND DANES—WHEAT AS A WILD PLANT—BARLEY—RYE—WHERE WAS THE VERY FIRST HARVEST?—VINE IN THE CAUCASUS—INDIANS SOWING CORN—EARLY WEEDS—WHERE DID WEEDS LIVE BEFORE CULTIVATION?—ARMIES OF WEEDS—THEIR CUNNING AND INGENUITY—GARDENERS' FEATS—THE IDEAL BEAN—DISEASED PINEAPPLES—RAISING BEETROOT AND CARROT—STORY OF THE TRAVELS OF SUGAR-CANE—INDIAN CUPID—BEETROOT AND NAPOLEON.

IT is difficult to understand the amount of labour and toil that has been spent on farmlands and pastures, if one only considers England.

It is often impossible to discover one square mile still covered by the natural wild plants. It is all under corn or arable, or rich artificial meadowland.

But from a Scotch hillside, as one looks down at the fertile valley below, one can see *first* where the mosaic of hedges and dykes stops, *then* where, after a narrow stretch of rough grass pasture, the cultivation ends; finally, where, ridge after ridge, rolling, heathery moorland, without enclosures and without any sign of man's handiwork, rises up to the highest peaks.

This fills one with a respect and reverence towards our forbears, which is increased by a study of corn, turnips, and potatoes.

Every one of these plants is a thoroughly unnatural, artificially bloated, and overfed sort of creature. Its constitution, as is usual with those who habitually overeat themselves, is delicate and unsound.

No cultivated plant could exist for more than a season if man did not look after it and protect it from its rivals and weeds. Moreover, they are a curiously assorted lot.

Wheat probably came from Asia Minor, Swedes from Scandinavia, Mangelwurzel from the Mediterranean, and Potatoes from Chile. Turnips and Carrots are indeed native Britishers, though the original wild carrot or turnip would never be recognized as such by any ordinary person.

The history of every one of them is interesting. The Oat is the true Teutonic and Scandinavian grain, which has more "fibre" than any other cereal. There is an interesting passage in Froissart's *Chronicles* describing the commissariat of those hardy Scotch borderers who raided and ruined the northern English counties whenever they felt inclined to do so. They lived for the most part on the cattle of their enemies, but each man carried a small sack of oatmeal and a griddle, or iron plate, on which to make oatcake. So that each man supported himself. His little rough pony also was quite able to look after itself.

That hardy plant, the Oat (*Avena sativa*) can be cultivated as far north as 69.50° N. lat. It is a native of Siberia and Western Europe.

It was oatmeal that supported the Norsemen who conquered Normandy and England, and who even dominated the Mediterranean. The Swedes of Gustavus Adolphus and the Danes of Canute also lived mainly upon oatmeal and porridge. It is true that in England oats are abandoned to the horses, but those horses are the best in the world. There can, of course, be no question as to whether the Scotch or English are the best!

The history of Wheat is a very complicated one; there are a great number of varieties and sub-species, all closely allied to our ordinary wheat, and difficult to distinguish from it. One variety occurs as a wild plant from Mesopotamia, near Ararat, over Servia, the Crimea, and as far as Thessaly, where entire hills are covered by it. This grain seems to have been cultivated at Troy, for Dr. Schliemann has found it at Hissarlik. It was, however, in cultivation long before the days of Achilles; it was grown by the Stone Age people, who lived in the lake dwellings of Switzerland. Another kind, "spelt" wheat, seems to have been the mainstay in ancient Egypt, in Greece, and all through the Roman Empire. It is now very rare, though it is still grown in Spain and in other countries where the soil is poor.

Grains of the true Wheat have been discovered in the Pyramids of Egypt, so that it also is very ancient. To-day Wheat extends to Norway (69° N. lat.), and may be grown up to 4400 feet on the Alps. India, United States, Russia, the Argentine, Chile, Australia, and many other countries, produce great crops of this useful and nourishing food. Its fibre is 3 per cent., albuminous matter 11-1/2 per cent., and carbo-hydrates 66·5 per cent. Oat has 10 per cent. fibre, 11-1/2 per cent. albuminous, and 57 per cent. carbo-hydrates.

One guess as to the origin of Wheat is that the first-named (Mesopotamian sort) is the original wild plant. By cultivation in the rich alluvial valleys of Mesopotamia and Egypt, improved kinds were formed. These have eventually replaced both "spelt" wheat and the wild race, but could only do so when richly-cultivated fields were ready for them. On poor soil and with bad cultivation, "spelt" is said to be even now the most profitable crop.

Wild Barley grows in Arabia and from Asia Minor to Baluchistan. It is very important in the colder regions of Northern Europe, in Tibet, and in China, but with us "John Barleycorn" is chiefly used for brewing.

Rye also comes from Asia Minor. It was not apparently known in Europe until the Bronze period, but is now "the chief cereal of the German and Slavonic nations." The black rye-bread is familiar to all who have travelled on the Continent. The straw is good fodder, and is used for making hats and for paper.

A very interesting point on which, however, it is quite impossible to come to a definite decision, may be noticed here. We will suppose what is quite as likely as any other theory, viz. that man as a gardening creature first settled somewhere in the Euphrates or Caucasian valleys.

What wild plants, then, would have been available for his experiments?

This particular region is an interesting and remarkable one. Most of our common British plants occur along the shore of the Black Sea to the Caucasus (apple, pear, nut, turnip, cabbage, carrot, and others, are all probably to be found there). On the Babylonian side of the mountains, there is a warm sub-tropical climate in which almost every useful plant can be grown. The desert also contains a few other valuable plants.

Near Ararat, Noah might have found rye, wheat, and barley growing wild. The Wild Vine also grows on the south of the Caucasus. "It grows there with the luxuriant wildness of a tropical creeper, clinging to tall trees and producing abundant fruit without pruning or cultivation." In that favoured district, the olive and the fig, the melon and cucumber, onions, garlic, and shallots, and other common garden and medicinal plants, can be found. Not far away is the native country of the camel, the ass, the horse, and most other domestic animals.

The buffaloes are puddling up the soil before the seed is planted.

Were these hillsides of Ararat or thereabouts, the first place where man sowed and reaped a harvest?

At any rate, in those flat, fertile, alluvial plains of the Euphrates, and also in Egypt, the first great cities arose.

But even in the later Stone Age, which may have been about 58,000 B.C., some of these Caucasian plants seem to have been in cultivation in Switzerland. Probably every subsequent invasion, first that of races with bronze weapons, and then of others in the Iron Age, brought with it new cultivated plants.

The Oat seems to be an exception to the rule, for, so far as one can gather, it was not a native of Asia Minor.

The first harvest was, however, in all probability, a very casual and occasional kind of thing.

Mason (*Origin of Inventions*) has described such a kind of cultivation which was in existence amongst the American Indians quite recently. "A company of Cocopa or Mohave or Pima women set forth to a rich and favoured spot on the side of a cañon or rocky steep. They are guarded by a sufficient number of men from capture or molestation. Each woman has a little bag of gourd seed, and when the company reach their destination she proceeds to plant the seeds one by one in a rich cranny or crevice where the roots may have opportunity to hold, the sun may shine in, and the vines with their fruit may swing down as from a trellis. The planters then go home and take no further notice of their vines until they return in the autumn to gather the gourds" (E. Palmer).

There is an interesting point about the cultivation of those early savage peoples who built up for themselves unhealthy but elaborate wooden dwellings in the Swiss lakes, in order to escape wild beasts and human beings who were even more dangerous and ferocious than they.

Weeds occurred in those cornfields, cultivated by stone implements, some 60,000 years ago.

The seed of an Italian weed had been introduced with their corn, and was discovered in Switzerland!

Weeds are an extremely interesting group. A proverb about the hardiness and multiplication of weeds can be discovered in almost every language. "Ill weeds grow apace," *Unkraut verbessert nicht*, and so on. They are very common. In fact weeds, wayside, and freshwater plants, have by far the widest distribution of all. There are twenty-five species which can be found over at least half the entire land surface of the earth, and more than a hundred occupy a third of it.

Moreover, many of our common weeds existed in Britain when the glaciers and ice melted away, and there were as yet no people able to cultivate the ground.

The Creeping Buttercup, Chickweed, Mint, Persicaria, Dock, and Sheep's Sorrel had already colonized the country, before the Great Ice Age came upon them, and at least fourteen weeds were here when the first corn-raising savages landed in Britain.

At first sight it is difficult to understand where and how they lived. One discovers a very few, however, if one botanizes very carefully along the seashore, or on river banks where landslips have occurred, and in other such places where bare ground exists which is not the result of cultivation.

There these weeds fulfil a very important and useful purpose. The "red smear" of a landslip is soon tinted green with Coltsfoot, Chickweed and the like, and the bare earth, which was useless and supported no green covering, is very soon made once more a part of the earth's fruitful field. In such places the weeds are soon overcome and suppressed by the regular woods, grass, or thicket of the district.

It is far otherwise in arable land, where man desires to keep the ground bare in order to give his own domestic plants the best part of the soil.

Let us look for a little at what actually happens in an ordinary cornfield. It is not merely one generation of weeds, but whole armies, that the farmer has to contend with.

When the young corn is growing up (1) the bright yellow Charlock grows much more rapidly, and the whole cornfield is golden with it. The Charlock grows to some eighteen inches high, flowers, and sets its seed before it is suppressed by the growth of the cornstalks, which, of course, may be three or four feet or more in height.

(2) Another series of weeds, such as Spurrey, are growing in the shelter of the tall stalks, and their flowers are ripened and their seed scattered long before the corn is cut. (3) Another series, such as Polygonums, etc., become ripe and are about the length of the corn, so that when it is cut and thrashed the seed of the Polygonum accompanies the grain and is probably sown with it. (4) Then there are such weeds as the False Oat grass, etc., which are taller than the Oat, and whose seeds are blown off and scattered all over the field before the harvest. One would think that those exhausted the series, but far from it: the farmer cuts and carries the crop, and for two or three days the ground is almost bare, but if you revisit the field a week afterwards you can no longer see the ground. The cut-off yellow stalks of the corn are set off by a dark continuous green carpet of flourishing weeds. This last, (5) the "waiting division" of the weeds, remain quietly until the corn is removed and then get through their flowering and seeding before the field is ploughed up or covered by grass.

Now if one thinks for a little over the cunning and ingenuity of these proceedings, it is obvious that each single weed has somehow learnt how to develop exactly at the right time. Those especially which are intended (by themselves) to form part of the seed mixtures must flower exactly at the same time as the corn. As a matter of fact, most seed mixtures are often full of weeds. In a single pound of clover seed, no less than 14,400 foreign seeds, including those of forty-four different weeds, have been discovered.

Others scattered on the ground will probably be buried and remain five to seven years below the surface, yet they are ready to come up flourishing as soon as they get a chance.

How has this been brought about? It is only since about 1780 to 1820 that our present system of farming has prevailed. In these 125 years, these weeds have found out exactly how to establish themselves.

The explanation is probably a very simple one. Every weed which did not bloom and seed exactly at the right time was killed and left no seed. This encouraged the others, who have gradually brought about the neat little arrangements above described. A process of selection has been at work. Those that would not modify their arrangements to suit new methods of farming have been suppressed.

But it is in some of the cultivated plants themselves that one sees the most extraordinary results of selection.

The Wild Cabbage is still to be found on sea-cliffs on the south-western coast of England, and the Wild Turnip occasionally occurs in fields. There is nothing particularly interesting or attractive about either of them.

Yet from the one has been produced cabbage, cauliflower, seakale, brussels sprouts, broccoli, and kohlrabi; and the other has given the endless varieties of turnips. For the most part these extraordinary changes have been brought about in a perfectly straightforward way, by just choosing the biggest and finest sorts for seed.

Some of the feats performed by gardeners in this way are almost incredible. A United States seedsman evolved the idea of a perfect bean from his inner consciousness. It had a particular shape which he described to a noted grower of beans. Two years later his ideal bean was produced!

The growers of pineapples used to have a great deal of difficulty on account of the pineapple cuttings becoming unhealthy. Sometimes 63 per cent. were more or less diseased. Then certain growers began to carefully select disease-proof pineapples, and finally reduced the percentage of diseased cuttings to four per cent. Another French observer (M. Roujon) by continually selecting the smallest seeds, was able to obtain corn only eight inches high.

But by far the most interesting and important researches have been those dealing with roots and tubers. Several people have, in fact, done in a few years what it took primitive man centuries to accomplish.

Thus, in 1890, E. v. Proskowetz obtained some seeds of the wild Sea-beetroot which is found on the south coast of France. By very careful selection he was able in the year 1894 to get good beetroots quite like the ordinary cultivated ones. These were biennials (not annuals like the wild plant), and had a large percentage of sugar—16·99 per cent. This was by selection in good and fertile soil. Vilmorin also obtained quite good carrots in the fourth generation by cultivating the wild form in rich and good soil, and selecting the best.

In fact there are in natural wild plants great differences between individuals, and when such plants are cultivated in good soil, where they have far more to eat than they require, the result is that they produce extraordinary and monstrous types.

These types are, however, more or less delicate, and are weak in constitution and easily killed. To prevent such variations those who wish to keep a race of seed pure are careful to keep it growing on poor land.

In 1596 the Hyacinth (*Hyacinthus orientalis*) was introduced from the Levant. In 1597 there were four varieties, and in 1629 eight kinds were known, but in 1768 two thousand forms of hyacinth were named and described.

Besides selection, the method of hybridizing or crossing is often used in order to obtain new or valuable strains. Generally both hybridizing and crossing are employed. This method has long been practised. Bradley, in 1717, writes as follows: "A curious person may by this knowledge produce much rare kinds of plants as have not yet been heard of"; and, in fact, peaches, potatoes, plums, strawberries, and savoys have all been greatly improved by hybridizing and selection. By crossing certain kinds of corn, such as the Chinese Oat and the wild European Oat, varieties have been produced by Messrs. Garton which at the Highland and Agricultural Society's trials produced 84, 87, and 99 bushels per acre, as compared with 58 bushels yielded by the ordinary Scotch Oat. With potatoes also astonishing results have been got.

The cart is being loaded up to carry the canes to the factory, where it will be crushed by the latest and most perfect machinery.

One single potato was sold for £50 not very long ago.

The Potato, like the Indian corn, tobacco, and a few other plants, is an inhabitant of the New World. Of other cultivated plants the native country is not known. No one knows where, for instance, Sugar-cane was first cultivated, but it has nine Sanskrit names, one of which, *khand,* is, or has probably at one time been familiar to us as sugar-candy. It was well-known when the Institutes of Manu were written, but that may have been somewhere between 2000 B.C. and A.D. 20.

One of the Hindu Indian deities, Kámadeva, who corresponds to Cupid, the God of Love, carries a bow made of sugar-cane, with a string which is composed of bees.

"He bends the luscious cane and twists the string
With bees: how sweet! but ah! how keen their sting,
He with five flowerets tips the ruthless darts
Which through five senses pierce enraptured hearts."

From India it seems to have been carried by Alexander the Great to Asia Minor, for it is mentioned by Herodotus. In the time of the Crusades it was discovered in Syria, and the Venetians learned something about it when the Crusaders returned to Europe. The Spaniards introduced the Sugar-cane to the Canary Islands in 1470. Then the Dutch took it to Brazil, and when they were expelled from that country by the Portuguese they transferred their canes to the West Indian Islands. Our English islands,

Barbados (1643) and Jamaica (1664), soon found the cultivation a very profitable undertaking.

The variations in price of sugar became in process of time of a very serious nature. In the year 1329 it is said that in Scotland a pound of sugar was worth one ounce of standard silver. But from 1780 to 1800 the price fell to 9d. The East Indian sugar began to compete with that from the West Indies about this time, but this was very soon crushed out by imposing a duty of £37 per cwt.

The West Indies were then very flourishing, but even before this the fatal word *beet-sugar* had already been heard. It was nothing at first but an interesting experiment by Professor Marcgraf in a German laboratory, who had extracted a little cane-sugar from beetroot in 1747. But in 1801 the beet was already in cultivation. Napoleon saw England's monopoly of the cane and judiciously encouraged the beet. The result of his far-seeing policy only became manifest a few years ago, for then the West Indian Islands, which we conquered and guarded against Napoleon at such fearful expense of blood and treasure, were almost worthless; Continental beet-sugar had ruined our colonial planters and our home refineries. It is in fact a most curious and interesting example of how a little judicious encouragement by a wise and far-seeing Government may destroy the profits of victory in a long, glorious, but yet ruinous war.

CHAPTER XXII. PLANTS AND ANTS.

MEANING OF PLANT LIFE—CAPTIVE AND DOMESTICATED GERMS—SOLOMON'S OBSERVATIONS DENIED BY BUFFON BUT CONFIRMED BY RECENT WRITERS—ANTS AS KEEPERS AND GERMINATORS OF CORN—ANT FIELDS—ANTS GROWING MUSHROOMS—LEAF-CUTTING ANTS—PLANTS WHICH ARE GUARDED BY INSECTS—THE AFRICAN BUSH—ANTS BOARDED BY ACACIAS AND BY IMBAUBA TREES—ANTS KEPT IN CHINA AND ITALY—COCKCHAFER *V.* ANT—SCALE INSECTS—A FUNGUS WHICH CATCHES WORMS.

THE world of plants supports all animal life, from the mite to the elephant. There are most intricate relations between one form of life and another. Thus a Rose tree attacked by an aphis or green-fly may be succoured by the slim ichneumon, or other thin-waisted fly, which lays its egg in that of the aphis. Another insect, say a spider, catches the ichneumon. A starling may eat the spider, and be itself eaten by an owl.

So that ichneumon and starling are friends to the Rose, whilst the other insect, the spider, and the owl are enemies. Yet both the starling and the spider are probably, almost certainly on the whole, friends of the Rose, although they are unfriendly in this special case.

With all other similar series or changes the final term is either a bird or animal of prey or mankind.

Until we introduce the idea of man as the culminating point of the series, the whole of it seems to be without any special meaning or advantage.

But when we think of how man utilizes the work of plants and animals, then the whole scheme becomes intelligible and complete; it is like a well-rounded story with a worthy and adequate end.

Moreover, what man has done so far is only an instalment of what he will probably succeed in doing. All who have brought up caterpillars or bees know that their greatest difficulty arises from certain minute insects or fungus enemies. We already know enough about these latter to fight them with some chance of success, but there are hundreds of other spores and germs floating in the atmosphere, and coming to rest on animals, on clothing, or on the leaves or petals of plants. These germs are now just as wild as, and infinitely more dangerous than the furious aurochs, the disdainful wild asses, or the ferocious wolves that our forefathers succeeded in domesticating.

Those bacteria, or germs, for instance, which are only one-thousandth of a millimetre long, are only visible by the help of a microscope. A row of three hundred thousand of them would be required to make an inch in length! Yet one of these germs can be mature and divide into two new germs in twenty minutes. In forty minutes there would be four, in an hour eight, and so on. The number after twenty-four hours is almost incredible.

These little germs stick to our clothes, fingers, lips, money, newspapers, and anything that is often handled. They hover in the air we breathe, permeate the food we eat, and inhabit water, and especially milk, in enormous numbers. Some of them are deadly. One might easily decimate a whole population, as indeed happened in the South Sea Islands when smallpox was introduced. Others are harmless and even necessary.

But to-day if you go into a bacteriological laboratory you will find hundreds and thousands of little glass tubes all neatly labelled and stoppered with cotton wool. If you read those labels you will see that the bacteria of all sorts of horrible and loathsome diseases have been captured and imprisoned. There is the deadly anthrax bacillus peacefully discolouring gelatine; in another, possibly the germs of hydrophobia may be undergoing a process of taming or treatment.

Each of these colonies of germs is under perfect control, and in many of them their natural wickedness has been so much alleviated that they are now useful aids to the doctor, who gives his patient a mild dose of the disease in order to accustom his system to resist accidental infection by the original type.

Yet what has been done already is only an earnest of what will no doubt be accomplished. Every farmer and ploughboy will in time sow his own bacteria; every dairymaid will make all sorts of cheese, from Camembert, Rochfort, to Gorgonzola, by sowing the right kind of germ upon it.

Man will no doubt cultivate the whole earth in the way that he now cultivates Europe and Great Britain, and will obtain mastery not only over his domesticated plants and animals, but over fungi, bacteria, and insects also.

Even if man had never risen above the state of the Banderlog of Mr. Kipling, there are other animals which cultivate and even combine together for warfare and conquest. In some respects they are better disciplined even than man himself, and they can defy all sorts of mankind except civilized man.

Possibly if man had not arisen on the scene, these insects might have developed some sort of civilization like that imagined by Mr. Wells in his story of the moon. We are only concerned with the relations of these ants to plants. Those who are interested in their conquests and civilization must consult the excellent account by Mr. Selous in his *Romance of the Insect World*.

The most interesting points about them are as follows. They gather a harvest and store it up for the winter. This habit of the ant was well known to the ancients, and is mentioned by Solomon. At the time of the French Encyclopædists, when the fashion of the times was all for destruction and disbelief, the fact that ants do so was ridiculed and flatly contradicted, and especially by the great naturalist Buffon. They pointed out that ants hibernated during the cold weather, and therefore required no food for the winter, so that Solomon's story was absolutely ridiculous.

For nearly a hundred years people forgot that Palestine and those other countries where the habits of ants had been reverently observed possessed a climate much too warm and mild to make the ants hibernate.

After careful study it has been discovered that the ants thoroughly understand the first stages of brewing!

The corn which they gather is not eaten by them in its hard winter condition. When taken into the winter nest of the ants this corn would very soon germinate and grow into a plant, but the ants manage to prevent this by some method which is not yet understood. If such a nest is left alone by the ants, the corn immediately begins to grow, but it is not allowed to do so till it is required for food. Should the store of corn get damp by heavy rain, or mould appear upon it, then the careful ants bring up their store into the sunlight and dry it there.

When it is required for food germination is permitted, but is soon stopped: the ants nibble off the growing rootlet of the seed. Then when the grain absorbs water and begins to change its starch into sugar, the ants suck in the sugar and reap the reward of all this labour and skill.

In the conduct of this germination of the grain they are, of course, far in advance of all the savage races of mankind.

There are certain South American species which go at least one step farther. They have their own fields—spaces three or four feet in diameter—which are entirely occupied by one single grass, the so-called Ant-rice (*Aristida stricta*). Dr. Lincecum states that the ants "work" these plantations very carefully, removing every weed or other plant that comes up, and sowing every year the new seed at the proper season.

These facts are sufficiently strange and startling, but there are even, apparently, species still more intelligent, who not only sow and reap, but actually prepare a soil and reap a crop of mushrooms, or at least, if not of mushrooms, of fungi. These wonderful little insects gather leaves and cut them into fragments of an appropriate size; they are then collected together so as to form a bed, and the fungus is introduced to this. The fungus is kept at a certain stage of growth by very careful treatment; the fruit-bearing ends are nibbled off, so that the young shoots come up indefinitely. The ants feed upon these fungus shoots, and get a crop indefinitely prolonged.

This is, of course, a system of agriculture far beyond that employed by any tribe of savages. Only man in a relatively advanced stage of agriculture grows mushrooms for himself. These facts, startling as they may seem, are apparently quite well authenticated and have not been seriously questioned.

There are a great number of leaf-cutting ants who are, indeed, amongst the most dangerous of the many insect pests in South America and elsewhere.

Wallace (*Revue Scientifique*, 179, p. 29), in speaking of the Saauba or leaf-cutters, describes how he placed a large heavy branch across the route of one of their columns.

The long line of laden ants was checked, and the greatest confusion set in at the head of the column. Each ant, for several feet down the column, then laid down its leaf, and all set to work to tunnel under the obstacle. This was managed in about half an hour's time, and the column then proceeded on its way.

Amongst other interesting and curious facts connected with these extraordinary insects is that some kinds are actually kept up by certain plants as a sort of standing army or police.

There are no less than 3030 species of plant which keep these standing armies of ferocious ants, or if they do not keep them, at any rate lay themselves out to attract them. The kinds which are attracted live upon sugar, and are strong, active, and extremely good fighters. When travelling through the bush in Africa, it is not unusual in some places to touch inadvertently one of these protected trees. In a moment one's hand and arm are covered by ants whose heads are dug deep down into the skin, biting with all their strength.

It is of course impossible to describe all the plants which protect themselves against injurious insects and even large animals in this way, but two of them must be mentioned.

There are certain Acacias which are particularly interesting. Like most of this order, they have large hollow spines instead of stipules at the base of the leaf. It is inside these spines that the troops of the police-insects live. These Acacias (Oxhorn Acacia, as well as *A. sphærocephala* and *A. spadicigera*) also produce *sugar*, which is secreted by peculiar gland-like organs on the stalks of the leaves, and even *albuminoids*, for at the tips of the leaflets there are peculiar little bodies which contain albuminous matter.

The Imbauba tree (*Cecropia spp.*) also possesses a standing army of these ants. It puts them up in the hollow pith in the centre of the tree, which is divided into large roomy spaces and makes a convenient nest. There is a minute opening by which they run in and out. On one occasion a naturalist found that the ants had been benumbed by a period of very cold weather, and in consequence had neglected their duty, and the trees had been stripped of their leaves by leaf-cutting kinds.

These last mentioned, the leaf-cutting ants, are especially dreaded by owners of plantations. Foreign or introduced plants are not specially guarded against their ravages by special secretions, as is the case with the native flora, so that the coffee and cocoa plantations are often severely injured. In some places man has copied those Acacias and Imbaubas, for in the orange plantations of the province of Canton, in China, ants' nests are collected and placed on the trees. Moreover, the different trees are connected together by bamboos, so that the ants can easily pass, as on a bridge, from one tree to another.

Near Mantua, in Italy, the same system seems to be adopted, and ants' nests are carefully placed near the fruit trees. Their use can be quite well understood, for Forel, in his work on the Ants of Switzerland, estimates that *one ants' nest will require* a supply of 100,000 insects a day during the season.

It is quite common to find ants crawling about on the outside of the large heads of the Garden Centaury and a few other Composites. If one looks carefully, one finds that there are streaks of honey to be seen coming from the scales. The honey is not produced in the flowers, and seems at first sight to be of no use at all so far as the plant is concerned, but that is very far from being the case. Here comes a cockchafer or other destructive beetle, intent on absolutely devouring and destroying the young flowers. At once the pugnacity and wrath of the ants are aroused. They take up a menacing and ferocious attitude, and the cockchafer passes to some other plant.

Such honey-glands found on the leaves and not connected in any way with the flowers, are more common than one would think. Even the common Bracken produces curious honey-secreting hairs when it is in a young condition. These attract ants which drive away caterpillars and other dangerous insect foes.

Many very dangerous insects are too small for birds, and can only be dealt with effectually by insects or fungi. Of these perhaps the most dangerous are the "scale" insects. The best-known one is very like a minute mussel shell. It is about one-quarter to one-third of an inch long, and can be sometimes found in quantities on apples; they are generally collected round the stalk. The mother insect has this scaly back, and lies down and dies on the top of her eggs, so that her scaly corpse forms a roof and a shield for her young ones. Like all pests of this sort, these creatures increase very rapidly.

A certain scale insect was doing an immense amount of harm in the orange plantations of Fiji, but it was destroyed by the introduction of lady-birds, and of a certain parasitic fly. It is said that these insects destroyed the "scale" in six months!

Experiments have also been tried with fungi. There are certain fungi which attack the bodies of living insects. So far, however, it cannot be said that the results have been at all satisfactory, for the propagation and infection of the living insects by fungus spores is not at all easy. There is also a certain feeling of doubt as to what may happen. Those fungi, and particularly bacteria, might set up dangerous epidemics.

Decaying meal contains hundreds of certain very curious worms called *Nematodes*. They are short, about one-twenty-fifth of an inch in length, and are smooth and very like minute eels. These creatures are very active, wriggling or swaying to and fro in a characteristic manner. Now in decaying meal there is a peculiar fungus. Like most fungi, it consists of very minute transparent threads which contain living matter or protoplasm. This particular fungus has branches, but also forms curious loops or belts. When one of these eel-worms is swaying about in the meal, it may quite well happen that its tail slips into one of these loops. If that happens, the fate of the worm is sealed, for the loop is elastic, and the more it wriggles the farther it slips in and the stronger it is held. The fungus then begins to grow, and forms a tube which grows *into the worm* and kills it. All the material in the worm's body goes to nourish the fungus. This extraordinary fungus has been described and figured by Professor Zopf, but seems to be a very unusual and rare form.

CHAPTER XXIII. THE PERIL OF INSECTS.

THE PHYLLOXERA—FRENCH SPORT—LIFE HISTORY OF THE PHYLLOXERA—COCKCHAFER GRUBS—WIREWORM—THE MISUNDERSTOOD CROWS—DANGEROUS SUCKLINGS OF GREENFLIES—"SWEAT OF HEAVEN" AND "SALIVA OF THE STARS"—A PARASITE OF A PARASITE OF A PARASITE—BUDS—THE APPLE-BLOSSOM WEEVIL—APPLE-SUCKER—THE CODLIN MOTH AND THE RIPENING APPLE—THE PEAR MIDGE—A CARELESS NATURALIST AND HIS PRESENT OF RARE EGGS—LEAF-MINERS—BIRDS WITHOUT A STAIN UPON THEIR CHARACTERS—BIRDS AND MAN—MOATS—DUST AND MITES—THE HOMES OF THE MITES—BUDS, INSECT EGGS, AND PARENT BIRDS FLOURISHING TOGETHER.

THE difficulty in describing the Romance of Plant Life does not arise from a want of romance, but the sieges, battles, and alarms are so difficult to see, and the enemies are so tiny, that the terrific contests continually going on escape our notice altogether.

When one does look carefully and closely at the life of a plant, one sometimes wonders how it manages to exist at all in the midst of so many and great dangers.

There are great swarms of insects which devour or burrow into it, or suck its life-juices. These are infinitely more dangerous than the relatively clumsy, heavy-footed, grazing animal.

Every part of a plant has its own special insect foe, and it is really difficult to understand how it can possibly escape.

Perhaps the "Achilles' heel" is the root, for, underground, plants get no help from the watchful and ever-present army of birds, who are, as we shall see, the natural police of the world.

The Phylloxera, for instance, which ruined the old and valuable vineyards in France, is a terrible little acarid, or mite, which attacks the roots. Too small to see, and impossible to kill without killing the plant, it laid waste the fertile hills and valleys of all South and Central France, causing millions of pounds damage. One reason for this destruction sprang from the universal sporting instinct innate in every Frenchman. Everybody goes out with his gun to destroy any lark, sparrow, or titmouse that is idiotic enough to remain in the country. Only birds can deal efficiently with insect pests. Take this horrible little Phylloxera, for instance; a single female in her life of forty-five days will lay about two hundred eggs. Each egg becomes a little grub, which after a few moments of uncertainty and agitation settles itself, and begins to suck steadily at any unoccupied part of the vine root. After ten to twelve days' life it will be laying eggs as rapidly as its mother. Thus in an ordinary summer the number of young ones produced from a single female becomes quite incalculable.

These pests are natives of America. Imported on American roots about 1868, they had in thirteen years practically ruined the vineyards in France, Spain, Portugal, Italy, and Germany.

All sorts of remedies were tried—saturation of the ground by poisons, flooding the vineyards to drown them, artificial cultivation of their insect and plant enemies, and many others.

The correct and satisfactory method has been at last discovered. American vines of sorts which are able to resist these Yankee mites have been imported, and the valuable French vines have been grafted on to them.

Another very dangerous root-enemy, which is common in this country, is the Cockchafer grub or Whitegrub. (But it is not nearly so bad as in France, where in the summer of 1889, a single farmer collected 2000 lb. of Cockchafers.) The grub (each female lays seventy eggs) burrows into the earth, and for no less than three summers remains below ground devouring indiscriminately the roots of everything he can discover. Underground, the mole is almost his only enemy, but the rooks, starlings, and gulls, which follow the plough, are watching for him. The Wireworm, Clickbeetle, or Skipjack, is also an underground demon which lives for three years, and gnaws and worries at plant roots for the whole of that time. It, however, shows itself above the surface.

A gentleman who had passed his whole life in the country complained, in my presence, of the damage done by rooks. He had had six thousand of them shot that summer, and remarked that he had seen with his own eyes one of them pulling out a young cabbage plant by the root. Of course it was quite unnecessary to point out that the poor bird was merely trying to get at the wireworms and devour them!

For some time I used to look out for great attacks of wireworm in turnip-fields: when one was recorded, I never failed to find that the crows had been ruthlessly shot down a season or two before.

All these, and many other insects, attack the roots, which would be, one would suppose, quite well protected in the depths of the earth. Therefore we find roots producing all sorts of poisonous substances, tannins, and even strong-smelling bodies, which keep off these pests.

It is perhaps the sucking battalions of the insect army which do the most harm. In themselves they are weak, stupid, and scarcely move from their birthplace. They live out their life wherever their long, lancet-like proboscis needles have pierced the plant's skin, but it is their power of multiplication that makes them really formidable.

Huxley calculated that if all the offspring of one "green-fly" lived, and if their broods also lived for ten generations, then the tenth brood of that original green-fly would contain more animal matter than the entire population of China. Green-fly would, as a matter of fact, go on increasing at this rate, were it not for the enormous number of enemies that prey upon them. A mathematical friend of Mr. Buckton calculated that in 300 days the produce of a single green-fly might be 210^{15}, that is 210 multiplied by 210, and then again by 210 up to 15 times!

In summer time one may often notice, especially on sycamores and lime trees, a peculiar shining, sticky, honey-like substance which covers the leaves. It is often so abundant as to drip like a rain of honey from the upper branches.

This "honey-dew" was a puzzle which greatly intrigued learned minds in the ancient world. Pliny speaks of it as the "sweat of heaven" or "saliva of the stars."

In reality, however, it is nothing but the excretions of hundreds of millions of these green-fly or aphides, which will be found established on the under side of the leaves, where, moored by their little anchoring talons and with their proboscis inserted in the fresh green leaf, they are sucking hard and steadily at the sugary juice. In twenty-four hours it was observed that a single individual gave forth forty-eight minute drops of honey.

Bees are very often tempted to collect this honey so abundantly produced, but this turns their own honey black, and may even make it poisonous.

Plants try to protect themselves against these pests chiefly by means of sticky or long hairs, by a thick skin, or by unpleasant tasting or smelling substances. But it is to insects such as lady-birds and others which devour the green-fly that they owe a deep debt of gratitude. In particular, there are certain parasitic insects which lay their eggs in their bodies. Not only so, but it is known that the eggs of some other insects are laid *in the egg of the green-fly,* and in one instance it has been found that yet another insect laid its egg in the egg of the parasite!

Some of the most interesting objects in nature are the buds in which, all neatly packed and stowed away, the young leaves and flowers remain awaiting the warm breath of spring. They are most interesting to examine: one finds series after series of overlapping scales which cover one another in the most ingenious way. No two are exactly alike, but each seems to have been moulded exactly to the proper shape. There is no waste anywhere, no useless expenditure of material. Very often turpentine or resin or a sticky gum seals up the joining of the scales. Every possible precaution seems to have been taken by nature. Neither rain nor snow can enter a winter bud. Neither can the cold of winter penetrate to the inside where the baby leaves and flower petals are cosily and tightly coiled up. But observe in the very earliest warm days of spring an extraordinary little insect, which has wakened up after its own winter sleep in the moss or lichen covering the rough and crannied bark of an old apple tree. This is the Apple-blossom Weevil, a beetle only about quarter of an inch in length, but with a curious snout or proboscis half the length of its body. This creature proceeds to the bud, and fixing its legs firmly, proceeds to bore a hole through the scales into the middle of the bud. She then places an egg inside, and goes on to put an egg in each of fourteen to forty-nine other buds. This takes a fortnight, and then she dies, probably satisfied that her duty is fully performed. A little footless, cream-white maggot develops in the apple-bud, which latter becomes rusty-coloured and dies away.

Another pest is the Apple-sucker, which lays her eggs in September on the fine hairs which cover the shoots. As soon as the weather becomes mild and warm, little grubs come out of these eggs; they are very small, and their bodies are almost flat. These tiny flat grubs, as soon as they are born, hurry off to the nearest buds and slip between their scales. They remain sucking the rich juices of the apple blossom until May or June, when they become perfect insects, and fly away so fat and well-nourished that they can live until September without feeding.

But those are by no means the only dangers. It is not till the apple blossom, which has escaped all those perils, opens in the spring time, after its petals have unfolded in the warm air and the young apple is already half formed, that the Codlin Moth begins to attack them. This tiny little moth is then extremely busy. She lays about fifty eggs, but only one on each young apple. It is put in the one weak spot of the apple, just at the top, in the base of the withered flower. The grub tunnels down to the core and feeds upon the seeds, which are entirely destroyed. When it has grown sufficiently, it drives another tunnel straight outwards to the skin. If the apple is still on the tree, the caterpillar lets itself down on a long silken thread and hurries off to hide in any convenient crack or crevice of the bark, or if the apple is already stored away, it conceals itself in the walls or in the flooring of the loft. The moths come out at the end of next May, just when the blossoms are getting ready for them. These codlin-moth apples cannot fail to have been noticed by the reader, as the tunnels in the ripe apple are most conspicuous. The gradual fattening of the caterpillar can also be traced, for its first tunnel down to the seeds is quite narrow, while the way out gets wider and wider as the creature became stouter and fatter whilst eating its way through the flesh.

The Pear Midge attacks at the same place, but the mother insect has a long egg-laying tube, and puts from fifteen to thirty eggs into the opening pear blossom. The pears go on growing, but of course are quite spoilt by the maggots within. These latter have a curious springing or jumping habit, and when they reach the soil bury themselves an inch or two below the surface.

So that all the care and neatness with which the young flowers and buds are packed up goes for nothing, and these insect pests get all the benefits of the apple and pear!

Besides these, there are hundreds of sorts of caterpillars which devour the leaves bodily. Cabbage-white butterflies, magpie-moths, gipsy-moths, diamondback-moths, and others, lay their eggs in hundreds. Many lay eggs each.

In the United States, somebody had sent an entomologist a present of some eggs of one of these moths. They were placed on a paper near a window which happened to be open; the entomologist went out, and the paper must have blown across the street into a garden on the other side. At any rate, two or three years afterwards it was found that some trees were badly attacked by this moth. Nobody thought much about this, though of course it was interesting to find a new moth. But the pest became a very serious one. In consequence of the stimulating air of the United States the moth multiplied with the most extraordinary rapidity, and it is said that about 300,000 dollars was spent in one year in the attempt to stamp it out.

All this happened because an entomologist forgot to lock up his eggs when he went away for half an hour!

These caterpillars and the locusts devour the leaves bodily, but there are others which live inside them. These so-called "leaf-miner" caterpillars make white irregularly-winding tunnels between the upper and the lower skin of the leaf. The tunnel increases or widens because the caterpillar itself grows fatter as it eats its tunnel. They can be seen on a great many leaves, and can be at once recognized by this peculiarity.

Plants cannot run away from their enemies like animals, and it would seem at first sight that their case was very hopeless. But it is not so, for there is a vast, active, keen-eyed, and eager army of helpers always ready for eggs and caterpillars.

It is birds that are of the greatest importance. A titmouse will eat 200,000 insects in a season. A starling has been seen to fetch food for its young ones from a grass paddock 100 yards away no less than eighteen times in a quarter of an hour. All the following are excellent birds, and without a stain upon their characters: the plover, partridge, robin, wagtail, starling. Crows and wood-pigeons are under suspicion, for though the latter do good in devouring the seeds of weeds, and the former in destroying wireworms, both are fond of corn and take large quantities of it.

Thrushes, mavises, and blackbirds are amongst the most persevering and useful of our friends, but they are certainly fond of fruit. Yet the good which they do is very much more than any possible harm which an injudicious indulgence in the juicy fruits of summer might bring about.

The sparrow cannot be given a character. Indeed, he is objectionable in every way, for he not only does no good himself, but he devours corn and drives away starlings and other valuable and interesting helpers.

But it is very difficult to say what will happen if man interferes with the regular working of Nature. The starling has been a pest in Australia, though here it does nothing but good work. We are still grossly ignorant of many simple but very important facts. Even when we do know something, as for instance, that the peewit's or plover's whole life is occupied in clearing the ground of wireworm, daddy-long-legs grub, insects' eggs, and the like, that does not help the bird in the least. Plovers' eggs are regularly sold in enormous quantities. Every farm-labourer collects them, and the farmer never dreams of interfering.

Man shoots down owls, kestrels, hawks, who prey upon mice, voles, and sparrows. Then, when some farmers are half ruined, he has Royal Commissions to find out why the voles have increased so much.

There are one or two peculiar contrivances found in plants which are intended to keep off insects, and which may be noticed here.

Thus, the importance of a moat (which almost always formed part of the defence of a medieval castle) had been already found out by one or two plants.

In a particular kind of Teazle and in a large Sunflower-like Composite (*Silphium laciniatum*) every pair of two opposite leaves run together, so that a little cup-like hollow is formed surrounding the stem, in which water collects. Insects climbing up the stem and trying to get at the heads of flowers fall in and get drowned in this water; their bodies may be seen floating about in it, and probably when these decay, their decay-products are of some use to the plant.

This curious contrivance is only a development of a very common arrangement. In most leaves you will find that rainwater is intended to run in a particular direction. There are little grooves and canals down which it is supposed to go, and dry, thirsty hairs may be found so arranged as to intercept part of it. Thus in summer the plants are not confined entirely to the water from the ground, but are also refreshed by the rain from above.

But if you look closely along these little channels, and especially at the base of the leaf where they join the stem, you will find that dust particles washed down by the rain collect and form little streaks and patches. The air is full of all sorts of dust particles which are made up of every conceivable substance. Many of these minute grains of dust will be dissolved in the water, and help to supply the plant with food. Nor is that all, for if you take a hand-lens and examine these dust particles very closely, you will very probably find small animalcula moving about. They are not pretty; in fact they are quite horrible to look at. These are tiny mites which live in these places. Their office is probably to eat up everything eatable (including eggs of insects and spores of fungi), and their excreta as well as their own bodies will probably be dissolved in the water and go to help the plants.

The most certain place to find them is on the leaves of the lime and other trees in August. On the under side of the leaf little bushes of hairs can be found just where the veins fork. It is necessary to take a pin and stir up these hairs to frighten them out, but when this has been done, the lens will show the disgusting-looking little creatures running hurriedly away. They are no doubt exceedingly annoyed at being disturbed in the midst of their sleep, for they come out and forage for anything eatable at night, retiring for the day into these hairy grottos. The structure of these grottos is very complicated. They are often like little caves with a narrow entrance, and the sleeping chamber is quite within the leaf.

A great many trees have these curious mite homes. The insects are generally the colour of the hairs, and are not easy to distinguish.

All those insects mentioned here have so arranged their life histories that they come into existence exactly at the proper season. The warmth of the sun, which opens the apple buds ever so slightly, stirs also the egg of the mite, the egg of the beetle, or the hibernating weevil, so that all these insect populations come into full active life just when they can do the most damage.

But one must not stop there; the bird population is also ready, and is building its nests and feeding its young, just so soon as the insect swarms are at their thickest and most dangerous stage.

Man walks clumsily through this intricate tangle of living plants and animals: he sets his big foot on a hedgehog (good for the insects), or on a mole (so much the better for wireworm), collects plovers' eggs (to the great help of every insect), shoots an owl (to the delight of voles and mice) or a whole brood of partridges, and in other ways makes a— — we had better say, shows that he is not so clever as he supposes himself to be.

CHAPTER XXIV. RUBBER, HEMP, AND OPIUM.

EFFECTS OF OPIUM—THE POPPY-PLANT AND ITS LATEX—WORK OF THE OPIUM-GATHERER—WHERE THE OPIUM POPPY IS GROWN—HASCHISCH OF THE COUNT OF MONTE CRISTO—HECKLING, SCOTCHING, AND RETTING—HEMPSEED AND BHANG—USERS OF HASCHISCH—USE OF INDIA-RUBBER—WHY PLANTS PRODUCE RUBBER—WITH THE INDIANS IN NICARAGUA—THE CONGO FREE STATE—SCARCITY OF RUBBER—COLUMBUS AND TORQUEMADA—MACINTOSH—GUTTA-PERCHA.

SUPPOSING that in China or Japan you meet a native who shows the following symptoms:—

(1) Eyes hollow and surrounded by a bluish margin; (2) pupils much dilated; (3) with a stupid appearance; (4) with an emaciated body; (5) of unsteady and staggering walk; (6) with a dreamy disposition;—then, you may be sure that he is an opium-smoker. In some of the Chinese provinces every man smokes ·03 to ·07 ounce of opium daily, but those who indulge to excess consume ·3 or even ·6 ounce. It is an excellent medicine when employed in a lawful and justifiable manner, for it calms the spirits and makes one sleep. But its use is *always* dangerous, even when employed in very small quantity, as in laudanum and morphia.

In the Fen country in England there used to be a very large sale of laudanum pills which keep off asthma and rheumatism, but even there it is a dangerous remedy, for it is only too easy to fall under the control of this drug either by injection of morphia, or by eating or smoking laudanum or morphia. De Quincey's *Confessions of an Opium-eater* and Kipling's *Gate of the Hundred Sorrows* give a lurid picture of the ruin of body and soul brought about by opium.

It is produced from the heads of the Opium Poppy (*Papaver somniferum*). Any poppy (or indeed any plant of the Poppy order) when scratched or wounded exudes a thick white or orange milky fluid. This is called "latex" (or milk); it is always more or less poisonous, and generally contains some sort of resinous matter. Thus when the plant is scratched or pierced, a drop of this milky latex comes out and at once hardens over the wound. Of course the plant is much benefited by this, for any destructive insect, unless it is a confirmed opium-eater, will be poisoned or killed; then also, if wounds are caused by wind, heavy rain, or animals passing, the scar is at once healed over and covered by the hardened opium, so that no dangerous fungus spores can get in to attack the plant. There is a mildew fungus and also a smut fungus (*Entyloma*) which attack the poppy, but both these enter by the stomata and live between the cells of the plant.

The general appearance of the Opium Poppy is quite familiar; its upright stems, large, clasping, bluish-green leaves and conspicuous flowers may be seen in many gardens. It is rather interesting, and in many ways; when young, the buds droop or hang down, and are entirely enclosed in two large green, hairy sepals. These last are soon thrown off, and then the flowers open out and display the petals with their rich black spots, and the crowded mass of stamens which surround the central greenish head. In bud these petals are "cramb'd up within the empalement by hundreds of little wrinkles or puckers as if three or four fine cambrick handkerchifs *sic* were thrust into one's pocket," as an old writer describes it (Grew).

Bees, and especially bumbles, are extremely fond of it, and even seem to be, in a way, opium-eaters, for they get quite exalted, almost intoxicated, and above their ordinary laborious selves. They scurry round and round the flower under the stamens or hover excitedly above it.

It is at this stage that the opium-gatherer begins his work; he goes round the beds and collects the petals of the poppy to use later on. The poppy-heads are then half grown and bluish-green, but they soon begin to turn yellow and ripen. When ripe they are most interesting to examine. There is a large platform covered by a radiating star-like ornament, which is the stigma. Underneath this is a circle of little holes just below the crown, but above the head. Each small hole has a flap. Now if you gather a ripe poppy-head on a fine dry day all these holes are open, and if you hold it upright and swing it vigorously from side to side the tiny seeds come flying out of the holes and will be thrown to a considerable distance. The stalk is supposed to swing in a high wind, and the seeds are really slung or thrown out of the holes. But if, when you come home, you put your poppy-head in water, or look at the plants in the garden on a very wet day, you will find that every hole closes or is shut up, because the small door mentioned above expands so as to close the opening.

The seeds are only sent out on a fine dry day; but they travel well. It was observed in America that certain poppies had been introduced as weeds at a certain place; in fifteen years they were found twenty-five miles farther on, so that they were colonizing the country at the rate of three-fifths of a mile per annum. The seeds themselves are very light and are of some value; they may be eaten like caraway-seed, as comfits, or crushed to supply an oil for lamps, or used as medicine. It is said that the value of the seed raised in France was in one year £170,000. The heads themselves are also valuable (they are worth 35s. per thousand), and even the dried stalks and leaves, for they may be used as fodder.

But the real reason why the plant is cultivated in so many parts of the earth is the great value of the opium obtained from it. This is gathered in the following curious way. As soon as the dew has dried off the plant, the cultivator goes round the beds and scratches every poppy-head with a tool made up of three knives tied together. That is the time recommended by Theophrastus, and it is apparently still the usual time to choose. In the late afternoon, from four to seven, he comes round again and scrapes off the congealed milk, which is then worked up into cakes and taken to the factory.

It is prepared by being kneaded, dried, and rubbed until it is of a pale golden colour. Finally, it is enclosed in a mass of poppy petals, sometimes mixed with the fruits of a kind of dock, and is then ready for export.

It is cultivated in a great many parts of the world—Turkey, Syria, Persia, France, China, the United States, Germany, Queensland, but especially in British India, where the immense plains at Malwa used to furnish opium worth about sixty million rupees annually (after deducting all expenses). This was mostly exported to China, and amounted to a tax of about threepence per head on every Chinaman; it was also sufficient to defray about one-sixteenth part of the expenses of our Indian Empire. The story of how Great Britain forced China to take our opium is not a creditable one nor agreeable to read. The plant was known in ancient Egypt, Persia, and Rome, and was used in China for at least two hundred years before our times.

Incisions may be seen in the bark of the tree. The rubber milk runs out from these into the vessel held in the man's hand.

What is supposed to be the original wild plant from which the opium poppy was derived seems to have been cultivated in the ancient Swiss lake dwellings, for the seeds of *Papaver setigerum* occur there in abundance. The price of the crop may amount to £90 or £120 per acre.

Another very ancient plant is the Hemp, *Cannabis sativa*. It was known to Herodotus, who says that "in the country of the Massagetæ there is a tree bearing a strange produce which they casting into a fire inhale its fumes on which they straightway become drunk." It is a tall, rather handsome annual, with stems from three to fifteen feet high. It is cultivated all over the world, from the Equator to 60° north latitude, but for different purposes. In India it is chiefly for the resin, "haschisch, churrus, bhang." (That was the drug used by the Count of Monte Cristo.) In Russia it is for the seed and the fibre that the plant is cultivated, and in France, Italy, and Austria the fibre seems to be the most important product.

Some of the plants produce only stamens or male flowers. The fibre given by these is stronger and more tenacious than that of the female plant, which, however, is finer and more supple. The fibre obtained from the cold northern districts of Russia is said to be the strongest of all.

The preparation of the fibre is a long, tedious, and laborious operation. It is also unhealthy, for the fibre has to be "retted" (steeped in water so that the soft parts decay), "scotched" (that is the hard wood must be broken and removed), and "heckled."

This last process is familiar to all who are interested in political matters. It consists of being drawn on hard points difficult to traverse and of a very fine and sharp character! Hemp is the commonest fibre for string, rope, etc.; it used to be employed for sailmaking by the Romans. Catherine de' Medici is said to have had two chemises made of hemp.

Hempseed is much appreciated by poultry and birds of all kinds (which makes both harvesting and sowing rather difficult); but the chief use of the seed is to furnish a fatty oil used for soft soap, lighting, and painting. The remains, after taking the oil, are employed as a cattle food, but it does not form a satisfactory cake.

The chief interest of hemp is, however, the drug that is made from the resinous juice. No doubt this has the effect of keeping off dangerous insects, for it is said that plants of hemp even keep off insects from other plants planted close beside them.

Sometimes the leaves and stalks are dried in order to make the drug "bhang." Many allusions to this substance are found in Eastern poetry, where it is called the "Leaf of Delusion," "Increaser of Pleasure," and "Cementer of Friendship," but madness is the result of addiction to its use.

The resin is collected by making the labourers put on leather aprons, and then run up and down vigorously through the hempfields. The resin is then scraped off the leather, or off their skins if they prefer to do without leather. It is either eaten or smoked. Burton describes how at every cottage door in East Africa the Arabs may be seen smoking bhang with or without tobacco. "It produces a violent cough ending in a kind of scream after a few long puffs." In small doses haschisch (resin) has pleasant effects, for people experience pleasant illusions, good appetite, excitement, and laughter, followed, however, after an interval by stupor and sleep.

People addicted to the use of haschisch roll their eyes violently, and have a wild, startled appearance.

Naturally so dangerous a drug cannot be recommended unless under the most exceptional circumstances, but it is employed in cases of asthma and insomnia. Haschisch and opium are the two great curses of the Chinese, Malays, and the inhabitants of British India and the East. They may be compared to "drink" in this country, but they are important medicines.

Among the most curious and interesting facts in Nature is the extraordinary variety of the ways in which at present gutta-percha and india-rubber are employed. We should not be able to ride bicycles, or in motor-cars; we could not use Atlantic cables and many electrical apparatus; our railway carriages would be most uncomfortable; golf would be impossible; we should have no waterproof coats and no goloshes sic, if it were not for these valuable and extraordinary substances, india-rubber or caoutchouc, and gutta-percha.

Their history is full of romance, but perhaps the most striking part of it is just this fact. Because a few (only a very few) plants found it necessary to protect their wood from burrowing beetles by a specially poisonous and elastic substance, *therefore* we can play golf and enjoy free-wheel bicycles.

The rubber is derived from the resinous latex or milky juice, which pours out from any wound in the bark of certain trees and creeping plants. This milk must be poisonous enough to kill the rash and intrusive mother beetle, who wishes to lay her eggs in the wood. It must be elastic, because the branches and stems swaying to and fro in the wind require a yielding, springy substance, but resin is contained in it, so that it promptly hardens and closes up the scar. The traveller Belt, in his *Naturalist in Nicaragua*, mentions that those trees which had been entirely drained of their rubber by the Indian gatherers were riddled by beetles, and in an unhealthy, dying condition.

Almost all the important rubber plants are found in wet, unhealthy, tropical forests; they are by far the most important jungle product in West Africa, as well as on the Congo River and in the Amazon valley.

It is quite impossible to describe the various rubber trees, and the different methods of gathering rubber, but it may be interesting to quote from an account of the method of its collection in Nicaragua, by Mr. Rowland W. Cater.

The best season for tapping the trees of *Castilloa elastica* is from August to February. It is best also to perform the operation early in the morning before the daily rain, "or in the evening after the rain has fallen. The milk ... is white and of the consistency of cream. The tree thrives best in moist but not marshy forests.

"It seeds in the tenth year, and ought not to be tapped before its eighth year, or its growth may be much retarded.

"On reaching the group of trees, which numbered seventeen of various sizes, my Carib friends first cut away the twining creepers that almost hid the trunks, and then carefully removed a couple of buruchas, natural ropes of rubber, formed in the following manner: From incisions in the bark, possibly caused by woodpeckers or some insect, the juice often exudes, trickling down the trunk, in and out of the encircling creepers, and sometimes reaching the ground. The milky stream coagulates and turns black as it runs, forming a long strip or cord, with which the huléros often tie up their bales.

"The parasites removed, Pete and José strapped on their espuelas (climbing spurs), fastened at the knee and ankle, and having dug a small pit or basin at the foot of each of a couple of trees, passed a ring of stout rope round the trunks and their own waists, and walked up with their machetes between their teeth. By lifting the rope at every step they were enabled to stand almost erect, and when lying back in the ring both hands were at liberty.

"José, whom I watched closely, commenced operations immediately below the first branch. With his broad-bladed sword he cut in the bark a horizontal canal which almost encircled the trunk and terminated in a V-shaped angle. From the point of the V downwards he next cut a perpendicular canal about two feet in length, which joined another horizontal channel ending in a V, and so on to the ground. In the last cut he inserted a large green leaf to serve as a funnel and guide the milk into the basin.

"The Brazilian rubber collectors always place a receptacle of tin or earthenware in the hole at the foot of the tree to prevent the admixture of grit or other foreign matters; they also strain the milk through coarse muslin; hence the greater value of Pará rubber. But Nicaraguan methods are primitive."

In the Congo Free State the taxes are paid by the collection of rubber. It is alleged that "if the demands for rubber or other produce were not satisfied, the people at fault were flogged often most barbarously with a thong of twisted hippopotamus hide, called the *chicotta*. Or else the natives were told to catch the women from the offending villages, who were brought to the *Chef de Poste* and imprisoned by him as hostages for the industry of their husbands. Or else the sentries shot some of the defaulters as examples to the rest. Frequently there were armed expeditions into refractory districts and widespread promiscuous slaughter. The cannibal soldiers of the State or of the Company sometimes feasting on the bodies of the slain."

The supply of rubber has of recent years shown signs of becoming exhausted. As time goes on the Indians of the Amazon and Orinoco must every year travel deeper into the inaccessible forests of the Amazon, Orinoco, or in Nicaragua. Every year also makes it more difficult for the Malagasy in Madagascar, or the Negroes in West Africa and the Congo, to gather sufficient rubber for the world's ever-growing needs. Liberia, the Negro Republic, is said still to possess plenty of rubber; but it is probable that the true solution of the difficulty will be found in the plantation of rubber trees. The exports from Madagascar in 1903 were valued at 2,585,000 francs; from Brazil, £9,700,000; from Nicaragua, 400,000 gold pesos (twelve pesos to the £); from the Congo, 47,000,000 francs; but even then about 85,000 rupees worth of rubber was exported from plantations in Ceylon. Unfortunately the trees do not begin to yield until they are eight years old, but the estimated profit per acre is very high, at least according to some authorities, who give a yield of £88 per acre (in Nicaragua).

One cannot help hoping that this will be the case. When one thinks, e.g., of the Uachins in the forests at the head of Namkong, who spend forty days in carrying their rubber on men's shoulders across the mountains to Assam, or of the horrible stories of the Congo Free State, plantation seems decidedly a more satisfactory method of supplying us with golf balls and bicycle tyres.

The first account of india-rubber is found in Herrera (Columbus's second voyage), who describes the way in which the natives play "with great dexterity and nimbleness." "They struck balls with any part of their bodies."

Juan de Torquemada in 1615 gives quite a good description of the Castilloa rubber:—

"The tree is held in great estimation, and grows in a hot country. It is not a very high tree: the leaves are round and of an ashy colour: it yields a white milky substance, thick and gummy and in great abundance. It is wounded with axe or cutlass, and from the wound the liquid drops into calabashes: Indians who have got no calabashes smear their bodies over with it (for nature is never without a resource), and when it becomes dry remove the whole incrustation."

The first patent for waterproofing seems to have been granted in 1791. A Charles Macintosh invented the garment named after him in 1823.

Very little of the commercial rubber is obtained from the common india-rubber Fig (*Ficus elasticus*) which we commonly grow indoors. This is one of those species of the Fig family which are generally found growing on the branches or trunks of other trees, though their own roots crawl down the trunk of the support to the ground. Once these roots have reached the ground, they take firm hold and grow so large and thick that they may be able to hold up the Fig tree even if the original support decays and crumbles away.

The gutta-percha which we use comes chiefly from Singapore, which is a sort of world's market for rubber. There are a great many different varieties and substitutes of this substance, but the best kinds come from Malaysia, Singapore, Sumatra, Java, and Borneo. The uses of gutta-percha and of vulcanite, which is manufactured from it, are very varied. Thus, it is employed for the soles of boots, door-handles, pipes, ear-trumpets, buckets, submarine cables, etc. It is indestructible in sea-water, and does not conduct electricity.

A very extraordinary exception to the general rule that latex is highly poisonous, is found in the famous Cow Tree of Venezuela. This tall tree (it is often 100 feet high) is found in large forests near Cariaco, on the coast of that country. Its milk is said to closely resemble ordinary milk in taste, and to be perfectly wholesome and nutritious, but it is rather sticky. This tree was responsible for all sorts of curious and extraordinary legends in the sixteenth and seventeenth centuries.

CHAPTER XXV. ON CLIMBING PLANTS.

ROBIN-RUN-THE-HEDGE—BRAMBLE BUSHES—CLIMBING ROSES—SPINY, WIRY STEMS OF SMILAX—THE WEAK YOUNG STEM OF A LIANE—THE WAY IN WHICH STEMS REVOLVE—THE HOP AND ITS LITTLE HARPOONS—A CLIMBING PALM—RAPIDITY OF TURNERS—THE EFFECT OF AMERICAN LIFE ON THEM—LIVING BRIDGES—ROPE BRIDGES IN INDIA—THE COMMON STITCHWORT—TENDRILS—THEIR BEHAVIOUR WHEN STROKED OR TICKLED—THEIR SENSIBILITY—THEIR GRASPING POWER—THE QUICKNESS WITH WHICH THEY CURVE AND THEIR SENSE OF WEIGHT—CHARLES DARWIN—REASONABLENESS OF PLANTS—CORKSCREW SPIRALS—THE PADS OF THE VIRGINIAN CREEPER—THE IVY—DOES IT DO HARM?—EMBRACING ROOTS—TREE IVY.

THERE are many plants which depend upon and cling to other more sturdy kinds, and which would be quite unable to live upon the earth at all if they had not developed the most beautiful methods of doing so.

In autumn, as soon as the leaves of the Hawthorn have fallen off, one is sure to find upon the hedges the common Robin-run-the-Hedge (Goosegrass, Cleavers, or Sticky Willie, for it is known by all these nicknames as well as by its proper name, *Galium aparine*).

Its stem is exceedingly weak, but it will be found sometimes to be six or seven feet long. It does not support itself, but is resting amongst and entangled in the outer twigs of the hedge in such a manner that it cannot be blown away by the wind or indeed picked out without its being broken. The young stems grow upright and are vigorous at first, but soon they cannot bear their own weight, and fall back upon a branch of the hedge. There are small curved little roughnesses along the stem and on the under side of the leaves of the Galium; these hitch on to the twig. Up to this point then the stem is supported, and the young part above grows until it also gets a lodgment, and so it goes on until it sometimes reaches right over the top of the hedge.

Its young flowering branches grow out towards the light away from the main stem, and the yellow withered stem in autumn rests upon the hedge just as a piece of string laid upon it might do.

The Bramble and Rose manage to get a support in very much the same way, but in Great Britain the Bramble generally grows in open ground and its branches take root.

The peculiar, curved-back prickles of the Bramble and its arching sideways growth would of course hang it on to any horizontal branches in the neighbourhood. Kerner measured the length of the stem of a Bramble which had interwoven itself into the boughs of a tree, and found that it was over twenty feet long, although it was only one-third of an inch thick. In Chile one often finds hedges of Brambles ten to fifteen feet in height, which have been formed by the aid of other plants, and also by the way in which the branches become entangled with one another.

Some Climbing Roses act in a very similar way, especially if grown on trellis, but the flower shoots always turn to the light like those of the Galium.

But it is the creepers and lianes of the tropical forests that are the most remarkable of all climbing plants. They twine round the stems and hang in great loops and grotesque folds from the branches. Sometimes in the dense shade it may be difficult to see the main stem, for it is quite thin, though as strong as a piece of steel wire. It often happens, when hurrying through a rather open part of the forest after game, that one's leg suddenly catches in a thin, spiny, wiry stem of Smilax or some such creeper. The first that one knows of the creeper is when a quarter of an inch of the spine is buried in one's flesh.

Away up amongst the branches and foliage far above one's head, leaves and flowers are developed on numerous branches which have vigorously pushed out as soon as they got near the sunlight, this tough, spiny, thread-like stem being their only connexion with the ground.

The development of these climbing plants is probably connected with the dense shade of forests. In such places a young stem growing up will become long and drawn out; its tip will droop over and hang downwards. But there is a curious peculiarity in the growth of all stems. The stem generally grows more rapidly at any one time on one side, say on the north, and therefore bends over to the opposite side. After a time it will be growing most rapidly on the eastern side and then its head points westwards, and so on. The result is that the tip of the stem swings in an irregular circle round the stem itself. Its head turns to every point of the compass in succession. Supposing a stone is tied to the end of a piece of string, and one swings the stone horizontally in a circle, then, if an upright stick is put in the ground and the string comes against it, the string will coil itself round the stick because the stone goes on swinging horizontally.

Our young climbing plant in the shade of the forest acts in exactly the same way. If there is any trunk of a suitable size, it will in the course of its revolving or sweeping round first touch and then coil itself round and round the trunk. Of these twining stems, one of the most interesting and beautiful is the common Hop. The young shoots or suckers which come from the ground may be seen waving their stems helplessly round in the air. If they cannot find something to cling to, then they form weak limp curves, but if one such shoot touches a pole it very soon obtains a hold, wraps itself round the support, and easily climbs up to a height of many feet. But the Hop is worth examining closely. If one passes the fingers along the stem, it feels rough and prickly. With the aid of a hand-lens, a whole series of most exquisite little hooks will be discovered. They are like small pimples with two or three very fine and minute, sharp grappling-hooks on the top. These prevent the stem from slipping off. It is also helped in climbing by its leaves, which curve outwards, and are also provided with grappling prickles on the under side. At the top of the stem the young leaves are close together, and folded near the point, so as not to interfere with the tip finding its way in and out of a trellis-work or amongst branches.

These grappling-hooks on the Hop are as perfect in their way, though by no means so beautiful and elegant as those which are found in the climbing palm, Desmoncus, so well described by Kerner in his *Natural History of Plants*. It is one of the rotang palms which reach lengths of 600 feet, though their stem may be no more than 1-1/3 to 2 inches thick. The leaflets towards the end of the leaf are transformed into strong spiny barbs which are exquisitely adapted to hang on to other plants. In many places, thickets in which these rotang palms have developed are so matted and tangled together that it is quite impossible even to cut into them, and they are practically impenetrable.

Some of our common British twiners climb very quickly. A complete turn round the supporting pole was made in England, at Charles Darwin's home, in the following times. The Hop took 2 hrs. 8 mins., Wistaria 2 hrs. 5 mins., Convolvulus 1 hr. 42 mins., and Phaseolus 1 hr. 54 mins. A Honeysuckle took 7 hrs. 30 mins. to make one complete turn round the support.

Recently Miss Elizabeth A. Simons timed the rate of growth of the same plants at the University of Pennsylvania. They seem to have been stimulated by the exhilarating atmosphere of the United States, for they were all growing faster. The Hop did its turn in 1 hr. 5 mins., Phaseolus took from 1 hr. to 1 hr. 20 mins., Convolvulus 57 mins. only, Lonicera from 1 hr. 43 mins. to 2 hrs. 48 mins., and Wistaria 2 hrs. But there are curious variations in the rate at which these plants revolve.

Thus when coming towards the light they go as fast as they can, but revolve more slowly, and as it were reluctantly, away from it. It has been found in one case that the shoot took thirty-five minutes to do the semicircle towards the light, and an hour and fifteen to twenty minutes going away from it, but this is not always the case, for sometimes the reverse takes place (Baranetzki).

These twining plants are not very common in Great Britain, and indeed in Europe. Some of them move or twine to the right (in the same direction as the hands of a watch or of the sun), such as Convolvulus (Bindweed), Phaseolus, Ipomœa, and Aristolochia. Others, like the Hop, Polygonum, Convolvulus, Honeysuckle, and Elephant's Foot, move in the opposite way from right to left, or "widder-shins." But there is nothing very important in this distinction, for the Bittersweet may be found twining in either direction, and in some plants part of a stem may be twining one way and the other in the opposite direction.

It is in the tropics, and especially in the rank, dark, moisture-laden atmosphere of the coast jungle forests, that these twiners attain their greatest development.

They show the most extraordinary variety. Sometimes a twiner hangs in elegant festoons from branch to branch, forming a convenient suspension bridge for monkeys. Sometimes four or five are wound round one another or twisted together, so that they look like some gigantic cable. In other cases they are knotted, looped, tangled, and twisted in the most inextricable manner.

Some creepers are flat, like green ribbons or broad bands. In others the dense mass of old, thick creepers and twiners round some sturdy trunk becomes so thick and so fused together that when the trunk dies the lattice-like arrangement of these creepers may keep them upright although the original supporting trunk is quite rotten and decayed away.

More usually, a tree will become unhealthy because its branches are overladen with the dense foliage and flowers of heavy lianes, and because both trunk and branches are so strangled in the embrace of great creepers that they cannot expand and develop in the proper way. Then a storm will overthrow the dead giant of the forest, and these creepers, entangled with all the surrounding trees, will produce ruin and destruction all around.

A regular duty of the foresters in India is to cut the stems of climbing plants. These twining, trailing, rope-like creepers are, in fact, natural ropes, and are used as such in India, Burma, and other places. Sometimes they form natural bridges of living plants extending across a stream. The great suspension bridges in the valleys of the Himalayas are sometimes made without a single nail or plank. They are just three ropes (one for the feet and two to hold on by) made of jungle creepers. Crossing one of these swinging, swaying creeper-bridges is not an easy matter for those whose heads are unaccustomed to depths of hundreds of feet below them, especially if combined with a motion of the creeper-bridge sufficient in itself to produce violent seasickness. Yet the natives run across them with loads on their heads!

But it is not necessary to go to the tropics to find interesting and ingenious climbing plants.

There is a very common little British plant, *Stellaria holostea* (the Star of Bethlehem, Great Starwort, or Stitchwort), which is common in shady places, light woods, and by hedges. In the spring it grows very quickly, and the pairs of leaves are shut together over the growing point, so that the end of the stem is narrow and can insert itself between the leaves and twigs of the neighbouring plants. As soon as such a growing end gets out of the foliage into the light, each pair of leaves opens out and curves backwards, making a pair of broad, curved hooks excellently suited to hang the stem on to the leaves or twigs. Then another period of growth follows, and again a new pair of hook-like leaves opens out. The stem may be five or six feet long.

In a rather rare Speedwell (*Veronica scutellata*) a very similar method is used, but the leaves have special little backward-pointing teeth on their edges which assist in the attachment process.

But these leaves are not to be compared as regards perfection of mechanism with the tendrils by means of which plants climb. These tendrils are thin, flexible, twining threads, which may be formed by the modification of whole leaves, in other cases of leaflets, or sometimes of branches. Sweet Peas, Vetches, Passion-flowers, Vines, and many other plants possess them.

They are like twining plants in the way in which they revolve or twine so as to wrap themselves round anything which they touch. They move much faster than twining plants. A Cobæa tendril only takes twenty-five minutes to make a complete turn, Passion-flowers take from half to three-quarters of an hour, and the Vine tendril takes a little over an hour to make one complete turn.

But in one way they differ altogether, for they are sensitive to contact. If tickled, they contract and embrace closely the object which is touching them. They show a most extraordinary sensibility and sensitiveness.

As a matter of fact, these tendrils have a finer sense of touch and a much more delicate feeling of weight than any human being. They detect the weight of twenty-seven inches of a spider's thread.

It is, however, best to explain what happens. A half-grown curved tendril of the Passion-flower is perhaps the most interesting to experiment with, but any sort of tendril does quite well. If one very gently rubs the inner or concave side of its little hook, then in a very few minutes, or even seconds, the tendril distinctly curves. If this has happened naturally, as when for instance it has been rubbing upon a pea-stick, this curve makes it curl round the stick, and the more it touches the more it curls, until the whole tendril is wrapped round the support.

It is, of course, quite impossible to explain it all exactly: the sensitive part on the inside of the curve differs from the outside or convex part of the tendril; the former has a layer of elongated, thin-walled cells, full of the living matter, protoplasm, which are absent on the outer side. Immediately the tendril touches the stick, the outer convex surface begins to grow rapidly. It grows from forty to 200 times as fast as the inner side which touches the stick! Very soon after it has clasped the stick the tendril becomes woody and forms a strong, woody, spiral coil.

These tendrils can be made to curve by a weight exceedingly small. The most sensitive part of our own skins is quite unable to distinguish so small a weight as is perceived by these tendrils. Even the sensation of taste can only be produced by a weight eight times as great as that shown by some of them. Tendrils curve very quickly after they have been touched. In twenty seconds some tendrils curve (*Cyclanthera*), others (*Passiflora*) take thirty seconds, and some of them require four to five minutes or even longer before they make up their minds to coil.

Even more remarkable, however, is the fact that they do not coil when raindrops fall on them, giving a much harder blow than small weights. If one tendril touches or rubs against another, it is said not to curve. They are persevering little things also, for Darwin got a passion-flower tendril to curve when struck or rubbed no less than twenty-one times during fifty-four hours.

If one reflects on all these curious facts, it is difficult to help feeling that these plants behave very much in the way that a reasonable animal would do. There are many other cases in which some vegetable does exactly what we should expect of reasonable beings under the circumstances. The tip of the root, the Sensitive Plant, the Monkey and Barberry flowers, are all well-known cases.

So that it is difficult to find anything in science to contradict the comfortable belief that wide-open flowers and stretched-out leaves of plants as they drink in the warm rays of the sunlight are really enjoying themselves, whilst they are doing their day's work.

All these interesting facts are so beautifully described and so carefully summed up by Charles Darwin, that we shall only earnestly recommend our readers to get first that fascinating book *The Power of Movement in Plants*, and then read all the rest of his works.

There are an extraordinary number of these plants and the tendrils are formed exactly where they will be most useful. Every part of a leaf may become a tendril. The whole leaf is changed into one in some kinds of Lathyrus. In a very beautiful creeper which is not so often grown in greenhouses as it might be (*Gloriosa superba*), the tip of the leaf only acts as a tendril. Leaflets are often made into tendrils. The Clematis is the most economical of them all, for the leaf-stalk coils round and forms little woody rings which hold up the plant.

Before leaving the subject of tendrils, it may be interesting to notice the queer corkscrew spirals in which they roll themselves up. These spirals are formed after the end of the tendril has tied itself to the support and become woody. The free part between the end and its own stem goes on revolving; now if you tie a piece of string at both ends and make it revolve, you will see at once that it must coil itself into a double spiral, one part in one direction and the other in the opposite way, with a flat piece between them.

One might be disposed to think no more about these double coils; but here comes in one of the curious, inexplicable coincidences which happen so often in plant life. Such a coil is much stronger than a straight bit of wire or string would be, because if pulled out it yields and is springy. That of course makes it less probable that the tendril will be broken. Attached by a series of wiry springs, the plant yields and sways to the wind, and it is not likely that it will be torn away. Besides this, the coiling of the tendril pulls the stem closer to its support, which is also a great advantage.

Certain Virginian Creepers and Vines behave in quite a different manner. The tendrils grow away from the light and so seek the shadow of the leaves. They are also divided into little branches. At the tip of each little branch is a small knob; if this should touch the wall or the trunk of a tree, etc., it immediately secretes a drop of cement and glues itself firmly to the wall. There is a curious difference in different sorts of *Ampelopsis* in this respect. There is no adhesive pad in one of them (*Ampelopsis hederacea*) until it touches, whilst *A. Veitchii* has them more or less ready for gluing before they touch (though they become much larger and better developed as soon as they rub against the wall).

One of the most interesting of our common climbers, "that rare old plant the Ivy green," has not yet been mentioned. It is exceedingly decorative on walls, especially on ruins and on old tree-trunks in winter time, where its dark, brilliant green is most effective.

A violent controversy rages as to whether it does good or harm. Unhappily it does not do any good to trees. It does not suck their sap, for its roots do not get through the bark, but it does choke, with its clinging branches, young tree-stems, and prevents their growing properly.

Also, in winter storms an ivy-covered tree is much more likely to be blown down. But on walls the ivy certainly does good, for it sucks up the moisture, and ivy-covered walls are much more dry inside than those which are exposed to rain.

Its method of climbing is very curious. All along the stem quantities of little roots are produced. They dislike light, like most roots, and creep into crevices and cracks, where they wedge themselves in by growing thicker. Thus the stem is anchored all along its length. It is curious to find that these roots are formed before a twig is actually touching the wall, so as to be ready for any emergencies.

One interesting little point in the growth of the ivy on a tree is perhaps worth mentioning. The main stem runs nearly straight up the trunk, and when young is pulled down into the crevices or cracks in the bark. But its branches leave the main stem at an angle of forty-five degrees or so to it; these latter may often grow in this direction for a foot or eighteen inches, but then they gradually begin to turn more and more distinctly up the tree.

Still these branches firmly clasp the trunk like arms spread out on either side of it, and make it almost impossible to dislodge the main stem.

Old plants of ivy entirely surround the trunk. The flowering branches grow straight out into the air, and have no tendency to cling to the bark. Their leaves are also different.

The ivy may be considered as a root-climber, although the branches assist by growing round the stem.

A curious instance has been given me of the longevity of ivy and its power of clinging to life. A correspondent mentions the case of a Scotch fir whose life was threatened by an ivy. The trunk of the ivy was sawn through. That did not kill it, at any rate immediately!

Probably the rain soaked up by the leaves, and by the roots in the crevices of the bark, kept it sufficiently fresh to cling to life. As it refused to die, a ladder was brought, and it was dragged off the tree. No doubt it would have died if the weather had been at all dry.

There are some very beautiful tropical plants which also climb by means of their roots. These roots, the so-called girdle roots, grow right round the stem and embrace it, so that the climber is perfectly supported.

It is impossible not to be impressed with the extraordinary variety of all these contrivances by which plants are able to escape the trouble of supporting themselves. But such ways of life involve certain disadvantages. Supposing there is nothing on which to climb, the stems trail feebly on the ground, and are probably soon choked by the surrounding grasses. Curiously enough, there are varieties of the Ivy, Wistaria, and the French Bean which are upright, and do not climb at all. The Tree Ivy has all its leaves like the leaves of the flowering shoot in the common form. In America, *Wistaria sinensis* is often grown as a standard tree, and does not send out the long shoots, sometimes thirty feet in length, which are common when it grows on walls. The dwarf French Bean has a thick stem and requires no support, yet it often puts out a long slender shoot which tries to twine round something.

In a tropical forest also, the creepers, though they damage the trees, yet manage to find space for their leaves and flowers: more vegetable matter is formed per square yard of ground than would be the case if there were no climbing plants.

CHAPTER XXVI. PLANTS WHICH PREY ON PLANTS.

THE KINDS OF CANNIBALS – BACTERIA – SPRING FLOWERS – PALE, GHOSTLY WOOD-FLOWERS – THEIR ALLIANCE WITH FUNGI – GOOSEBERRIES GROWING ON TREES – ORCHID-HUNTING – THE LIFE OF AN ORCHID – THE MISTLETOE – BALDER THE BEAUTIFUL – DRUIDS – MISTLETOE AS A REMEDY – ITS PARASITIC ROOTS – THE TREES IT PREFERS – THE *CACTUS LORANTHUS* – YELLOW RATTLE AND EYEBRIGHT, OR MILK-THIEF, AND THEIR ROOT-SUCKERS – BROOMRAPE AND TOOTHWORT – THEIR COLOUR AND TASTES – THE SCALES OF THE TOOTHWORT WHICH CATCH ANIMALCULA – SIR STAMFORD RAFFLES – A FLOWER A YARD ACROSS – THE DODDER – ITS TWINING STEM AND SUCKER-ROOTS – PARASITES RARE, DEGENERATE AND DANGEROUSLY SITUATED.

THE word *cannibal* is often used in a very loose and unscientific way. Amongst some savage tribes it is the custom to eat old people and young children; but this is only in seasons of famine and scarcity, when there is no other food available, and not because they are specially fond of them. But amongst other tribes wars are made for the special purpose of capturing fat young people to cook. Sometimes they have become so accustomed to such delicacies that they are unable to get their food in any other way. Of course, when tribes become "pure cannibals" of this last type they have to be destroyed like wild beasts.

Among plants we find all sorts of transitions and degrees of cannibalism. There are plants which sometimes, and, as it were, accidentally, attack others. But there are also real cannibal plants which live entirely on the life-juices and sap of other plants, and cannot exist by their own labours at all. Moreover, we can find almost every conceivable state of transition. These can be clearly and definitely traced from those plants which depend on the labour of their own roots and leaves to others which have no leaves, and which consist merely of one large flower and a large adhesive sucker fixed on some one else's root.

The difficulty is very often to know where to draw the line. Probably no flowering plant is quite independent of the labour and work of its neighbours. As we have tried to show in another chapter, a long preliminary cultivation by bacteria, lichens, and mosses is required before flowering plants can develop on bare rock. That is also necessary in all cases where the soil is mineral or *inorganic*, without any *organic* dust or fragments of vegetable or animal matter. Bacteria must always begin the work by preparing nitrates and other salts.

So that only those bacteria which weather rocks can be called really free and independent. But other bacteria, such as those which cause typhoid, anthrax, hydrophobia, etc., are the best possible examples of pure cannibals, or, as they are usually described, parasites.

This last word is derived from a peculiar class of people in ancient classical times, who used to appear whenever a meal was going to begin, and received food without giving anything in return. They are represented by our tramps or by the "sundowners" in Australia, who appear as soon as the evening meal is ready and when there is no possibility of going any further on their journey.

The way in which plants became parasites or cannibals is a very interesting part of plant life, and we shall try to trace some of the various stages.

To begin with, if one looks out for them in spring one is sure to find a whole series of beautiful spring flowers. There is the Primrose, with its bright, hardy, yellow flowers; the Violet, whose strong perfume much annoys the huntsman, for it spoils the "scent" and shows him that the end of winter has come; the delicate little Moschatel, the Lesser Celandine, the Bluebell or Hyacinth, Dog's Mercury, the Male and the Lady Fern, and many others.

Most of these begin to grow and are in flower early in the season. That is because they are living on the *dead leaves* of the last year, or rather of two or three years ago. Their roots are breaking up and devouring, with the help of worms, beetles, and insects, the leaf-mould of past seasons.

They are quite dependent on the trees; they cannot exist except where such leaf-mould is formed.

But it is very difficult to tell whether these humble little herbs which live on the scraps that fall from the tall trees are either parasites or clients, which last do some good in return for their share.

Probably they are distinctly useful and good for the forest if this is considered as a whole *establishment*. They use light which would otherwise be wasted, and their own dead leaves increase the annual deposit of leaf-mould.

There are other plants, such as the Bird's-nest (*Neottia*) and Coralroot Orchids, as well as Monotropa and others, which also live on the rich, decaying leaf-mould of forests, but these are generally pale in colour, for they possess but little green chlorophyll. They are more directly dependent on the mould and have ceased to do much work for themselves. Most of them in fact have entered into an alliance with fungi, and use these fungi to get their food material from the dead leaves.

Such fungi are always abundant in good, well-grown forests. These Orchids and Monotropa have their roots and underground stems covered and wrapped round by the fungus threads, which extend from them in every direction, breaking up and decomposing the dead leaves.

The colour of Monotropa is a pale waxen yellow, that of the others is usually a ghostly pale, opalescent, steel-blue or coral-like hue, which makes them very distinct in the dim, mysterious shades of the forest.

These plants are undoubtedly of use, for they break up and decompose the leaf-mould.

Another very interesting group are not well represented in this country. Sometimes one may see on an old tree a Gooseberry bush in full foliage quite high up the trunk in the fork of the branches. In sheltered woody ravines, Polypody ferns are often established on old moss-clad branches, where their green fronds hang over to catch as much as they can of the sunlight. But Orchids, Bromeliads, and Ferns which grow upon the branches of great trees are one of the most conspicuous and beautiful features of tropical woods. It is for these tree-orchids that the orchid-hunter braves the head-hunters of Borneo or traverses the precipices and rugged forests of Guatemala and Brazil. It is often necessary to cut down a tall tree in order to get the orchids in its higher branches. Often, however, this is unsuccessful, for the tree is so held up by creepers and other giants of the forest that it never reaches the ground!

Then, after being stripped from the branches, in some out-of-the-way forest-clad range of Burma, Celebes, South America, or Madagascar, these orchids are dried, put up in crates and packed off to London, where they are carefully cultivated in hot-houses and persuaded to flower. They may be worth sixpence or they may be worth £500 each, but no one can tell until they have flowered in London.

But the romance of the orchid-hunter is not exactly what we have to describe here. It is rather the romance of the life of the orchid itself.

It is perched high up on the branches of the tallest trees in the forest, exposed to sun, exposed to wind, and quite unable to gather either salts or rain from the soil. How, then, does it manage to live?

These orchids, it must be remembered, are only found in out-of-the-way and feverish, unhealthy places, where the aboriginal savages still lurk and endure a dreadful existence of hunger and starvation in dense tropical forests.

Now the word "dense" explains the whole story. Those forests are so thick, so full of giant trees and exuberant growth, that civilized man even to-day in 1906 can make nothing of them, and leaves them to the savage. The reason why vegetation is so luxuriant is simply that there are both plentiful moisture and a hot, tropical sun. That makes the life of the orchid possible, and also ensures malaria for the hunter.

It hangs out into the moist air long pendulous roots which act as so many sponges absorbing and soaking in moisture. The tremendous energy of growth covers bark and branches with creeping plants innumerable, with a profusion of moss, liverworts, and ferns such as we cannot imagine from our own experiences in this country. So the roots of our orchid find on the branches rich leaf-mould, and it lives happily and contentedly on the salts and moisture accumulated by the mosses and other plants. Its leaves are fleshy and succulent, rather like those of a desert plant, so that it can store up water against a season of drought.

These plants which grow in this way on other plants, do not, as a rule, greatly injure them, but many have not stopped at this stage. Take, for instance, the Gooseberry growing in the fork of an old tree. Some bird has been eating gooseberries and dropped the seed there. The roots of the gooseberry will grow down into the rotten part of the trunk. Earth and leaf-mould will accumulate there, and it is quite probable that the whole inside of the tree will decay away. The roots of the gooseberry will, if only indirectly, help in this decay.

But it is far otherwise with another set of plants—the Mistletoe and its allies. There is plenty of romance connected with the mistletoe. Dr. M. T. Masters says as follows: "The origin of the modern custom connected with mistletoe is not very clear. Like many other customs, its original significance is only guessed at. If known, perhaps, the innocent merriment now associated with the plant would be exchanged for a feeling of stern disapproval, and the mistletoe would be banished from our homes. In such a case ignorance is bliss."

It will be remembered that all the gods of Iceland were once gathered together so that a general oath might be exacted of every plant "that grew upon the earth," that they would do no harm to Balder the Beautiful. The Mistletoe did not take the oath, because it does not grow upon the earth but upon a tree. Then the enemy fashioned an arrow out of the mistletoe, and killed Balder. There is a modern idea that the story is a myth representing the death of Spring, for a great many similar stories occur in widely distant places.

However, it seems pretty certain that the plant was a sacred one to the Druids in the time of the Romans.

Ovid speaks of this in the line, "Ad Viscum Druidæ cantare solebant." At their solemn meetings, which were held in remote sacred groves, a Druid clad in white robes cut the mistletoe with a golden sickle. Then, apparently, human sacrifices were offered and a general festival took place.

Some remnant of this custom seems to have persisted in Herefordshire until recent times, for the tune "Hey derry down, down down derry" (which means *in a circle move we round the oak*) is supposed to be a relic of the hymn chanted by the Druids when they had found mistletoe on the oak.

It was said in the Middle Ages to be a useful cure for apoplexy, madness, and giddiness. That is not at present the general view. Indeed, under present conditions it might conceivably promote the last and even the second of these disorders, though in an agreeable way!

The Mistletoe and its allies, Loranthus and Arceuthobium, grow upon the branches of trees like the orchids and gooseberries already mentioned, but they differ altogether in having a special kind of absorbing root which sinks down into the bark until it reaches the wood of the "host" tree. The sap running up the tree is then tapped by this root, and goes to supply the mistletoe with water and salts in solution. It has, however, its own green leaves. Thrushes eat the berries of the mistletoe; they will be left upon a branch with the *guano*; as the latter dries up, the seed is drawn to the underside of the branch, and sticks in a crack or crevice; it then sends the sinker-root mentioned above into the branch.

Every year afterwards new mistletoe "roots" are formed which grow through the soft part of the bark and send down sinkers into the wood. Cases of Mistletoes forty years old have been recorded. The trees which they prefer are the Apple, and after that Black Poplar, though mistletoe may be found on Silver Fir, various Pines, and others. It is more difficult to get it to grow on the Oak than on any other tree. Indeed, only seven cases of mistletoe growing on oak have been recorded in this country. It is quite a valuable crop in some places, and is sent in tons to the London market.

There are many species of Mistletoe, and at least one kind attacks, and is parasitic upon, another species of Mistletoe.

Most Mistletoes and Loranthus have their own green leaves, and only take from the plant to which they are attached sap and mineral salts. But in Chile there is a beautiful Loranthus that has practically no green leaves at all. Its blood-red flowers grow in dense masses upon the giant Cactus, which is common on the drier hills, and these are always mistaken for the Cactus's own flowers, which are quite different. These almost leafless Loranthus, and the curious Arceuthobium are more parasitic than ordinary mistletoes, for they obviously take other food material (probably sugar and albuminoids) from their "host."

Another series of parasites or cannibals are quite common in Great Britain. One often sees in some meadow that the grasses are growing in a scanty and unhealthy manner; one then notices amongst them numbers of the Yellow Rattle or the Eyebright (which the Germans call *Milk-thief*). These plants are not very remarkable in any way, but if one examines them closely one sees that the leaves and stems are more purplish-red than is at all usual with our ordinary flowering plants. But if you dig up some specimens very carefully, then the wickedness of the Yellow Rattle and Eyebright becomes apparent; every here and there upon their roots are little whitish swellings which are firmly attached to the roots of other plants (generally of grasses). These two robber plants send from these swellings minute sucker-roots which pierce into the grass-root and intercept the water which the grass has been absorbing for itself.

They are therefore parasites, and indeed they may cause a considerable loss of forage in a meadow.

A good many other British plants are root thieves. Besides these two, there are the Cow-wheat, Red Rattles, Toadflax, Broomrapes, and Toothwort.

A curious point about them is that they differ amongst themselves in the degree in which they are dependent on the work of others. Some are able to grow quite well without any such extraneous help, but the Broomrape and Toothwort are entirely dependent on others' labours. They have extremely little chlorophyll and very small leaves, and are clearly parasites "pure and simple."

There are about 180 species of Broomrape (*Orobanche*). All of them attack roots, and most confine their attentions to one particular flowering plant. Their colours are generally very striking and unusual. Our British species are reddish, flesh-coloured, or dirty white, but some of the foreign kinds are blue or violet, yellow, or yellowish to dark brown. Generally the seedling Broomrape worms its way down into the earth till its root-tip touches the root of its special favourite host, then the root of the Broomrape fixes itself for life; its suckers grow into the host and absorb all the food material which it requires. Those kinds which attack Tobacco and Hemp are dangerous pests and do considerable damage.

The Toothwort (*Lathraea*) is so called because its scales have a sort of resemblance to human teeth. With the curious superstition which prevailed in medieval times, it was supposed that the plant must be a remedy for toothache because it resembled teeth. Unfortunately this is not the case.

It is, generally, quite like the Broomrape in its method of growth, but it sends out long thread-like branching roots with suckers on the ends, which become fastened on the Hazel roots. For several years the plant remains underground and forms very odd-looking, white, scaly branches. These scales are rolled back in such a way as to form peculiar and irregular cavities which open to the outside near the tip of the leaf. There is no doubt that animalcula of sorts get into these cavities and probably die there. In that case, their remains will form a useful supplement to the diet of the plant. The following remarks, however, taken from Kerner have been disputed by other botanists.

Certain of the cells lining these cavities "appear to send out delicate filaments.

"When small animals penetrate into the labyrinthine chambers of a Lathraea leaf and touch the organs just described, the protoplasmic filaments are protruded and lay themselves upon the intruders. They act as prehensile arms in holding the smaller prey, chiefly Infusoria, and impede the motion of larger animals so as to cut off their retreat. No special secretion has been observed to be exuded in the foliar chambers of Lathraea. But seeing that some time after the creatures have entered the chambers, the only remains of them that one meets with are claws, legs, bristles, and little amorphous lumps, their sarcode-flesh and blood having vanished and left no trace, we must suppose that the absorption of nutriment from the dead prey here ensues...."

But strange as these Broomrapes and Toothworts may be, they are quite inconspicuous as compared with the gigantic parasites found in Sumatra and Java.

In 1818, when Sir Stamford Raffles was making a tour in the interior of Sumatra, his party came across one of those extraordinary plants which have been called after him.

Imagine a gigantic flower in shape resembling a very fleshy forget-me-not, but more than a yard across! The colour is a livid, fleshy tint, and the smell is like that of a charnel-house. This extraordinary *Rafflesia Arnoldii* is the biggest flower in the world. It has no proper stems or leaves, but consists merely of this huge flower-bud attached to the roots of Figs, etc., which traverse the ground in these forests. It is said to be only found in places frequented by elephants, which are supposed to carry its seeds on their feet.

There are four other kinds known: all of them occur in Sumatra, Java, and other neighbouring islands. *R. Padma* for example, has a flower about eighteen inches across. The central part is a dirty blood-red, while the lobes have almost the colour of the human skin. This also has a "cadaverous smell, anything but pleasant."

These weird Rafflesias seated on the roots "which wind about on the dark forest ground" have impressed every observer.

Yet if one glances back, it is interesting to see how insensible are the transitional steps which lead from independent life by the plant's own exertions to these last "pure parasites," which are entirely dependent on other plants for everything that they require.

The only other flowering plant which we shall mention in this chapter is now fortunately very rare in Great Britain. This is the Dodder, *Cuscuta.* It belongs to the Convolvulus or Bindweed order, but is entirely different from the rest of the family. Some climbing plants do throttle or choke the trunks of young trees if they twine round them too closely, but the Dodder has an entirely special and peculiar way of supporting itself to the detriment of others. It has no roots, no leaves, and scarcely any green chlorophyll; the Dodder is just a twining, thread-like, yellowish stem which carries here and there small round clusters of little convolvulus-like flowers. Wherever the Dodder thread twines round a hop or other plant, it puts out small suckers which drive their way into the stem of the hop and take from it all the food which the Dodder requires. When well developed it forms dense yellowish tangles of intricately entwined threads, which may cover whole bushes and entirely destroy the supporting plants. The Flax, Clover, and Hop Dodders are perhaps the worst of them all.

There are some rather interesting points in the history of the tiny dodder-seedling. It remains, quietly waiting, for about a month after most other plants have germinated.

Then it begins to grow rapidly: its tip pierces the soil and becomes fixed in it; then the rest of the little thread-like seedling begins to curve round or revolve. If it touches a grass or even a nettle stem, it twines itself or coils round it, drives in its suckers, and, on the strength of the nourishment which it extracts, it goes on revolving or turning until it forms the dense tangled masses referred to.

Then an eruption of flowers appears, from which later on hundreds of tiny seeds are let loose which will become Dodders in their turn.

The series of parasitic plants which have now been mentioned form a very interesting set. It must be pointed out that those which live merely on dead vegetable matter are "good" plants. They help on the quick and thorough employment of worn-out material.

Nor can we say off-hand that other parasites are "bad." They do kill other plants and do them harm, but then, are they not like a cattle-breeder who sends his inferior cattle to the butcher, keeping only those which are the very best of their kind? Perhaps these plants, by destroying the weak and unhealthy kinds, are doing a great deal of good.

Another interesting point about such parasites is that they are generally *rare*. They must be less common than their "host." Yet another is that they are all "degenerates." They show distinct traces of decay and bad development in their flowers and seed. That is also true in the case of parasitic animals.

Whether they do good or harm to the world of plants is doubtful, but there is no doubt that they are doing harm to their own chances!

CHAPTER XXVII. PLANTS ATTACKING ANIMALS.

BRITTLE STAR *V.* ALGÆ—FUNGUS *V.* MEAL-WORM—STAG-HEADED CATERPILLARS—LIVERWORT *V.* SMALL INSECTS—NATURAL FLOWER-POTS—WATERCUPS OF BROMELIADS—SARRACENIA AND INQUIRING INSECTS—AN UNFORTUNATE CENTIPEDE—PITCHER-PLANTS: THEIR CRAFTY CONTRIVANCES—BLOWFLIES DEFY THEM AND SPIDERS ROB THEM—BLADDERWORT'S TRAPS WHICH CATCH SMALL FRY—HAIRS AND THEIR USES—PLANTS USED AS FLY-PAPERS—BUTTERWORT *V.* MIDGES—ITS USE AS RENNET—SUNDEW AND ITS SENSITIVE TENTACLES—PINNING DOWN AN INSECT—SUFFOCATING AND CHLOROFORMING THE SUNDEW—VENUS' FLY-TRAP WHICH ACTS LIKE A RAT-TRAP—HAVE PLANTS A NERVOUS SYSTEM?

ON the whole the animal world preys upon the vegetable world, and is in a way parasitic upon it. Indeed, the connexion between the two is very intimate—that of the diner and his dinner. One can scarcely imagine a more intimate connexion than this!

There are, however, a great many cases in which plants have turned the tables on their enemies and deliberately laid themselves out to catch and to destroy, to feed upon and to devour insects and small animals. One finds a few examples in almost every group of plants.

Thus there are certain green seaweeds or algæ which are said to attack and prey upon those peculiar sea-urchins known as Brittle Stars. The fungus which forms loops, acting exactly like a poacher's rabbit-snare, in order to catch mealworms, has been already mentioned.

Sometimes in the summer one may notice a little red club about two to three inches long sticking out of short grass. If one carefully pulls this up it is found to be growing out of a dead chrysalis or grub. It is a fungus whose spores have attacked the caterpillar; they have developed inside its body, and eventually, having completely eaten up the insect, form the red club, which is producing hundreds of thousands of spores intended to attack other caterpillars.

The branches like stag's horns are the fruit of a fungus, Cordyceps Taylori, which lived inside and killed the caterpillar.

An allied fungus forms a peculiar branched fruit rather like a minute stag's horn, and the caterpillar may be seen for some time crawling about with this extraordinary fungus sticking out of its head. Of course the bacteria are, some of them, by far the most dangerous foes of animals.

Then there is a small Liverwort, a little red, moss-like plant (*Frullania tamarisci*), which may be found growing on the bark of trees, which is said to catch animalcula in the small sack-like leaves which are underneath the ordinary ones.

But it is amongst the higher flowering plants that one discovers the most extraordinary and purposeful arrangements for capturing and digesting insects and other creatures.

In the case of many of these insectivorous plants, traps or pitfalls are prepared for the insect to fall into.

There are many plants in which the rain is intended to run in one particular direction, and it is not at all uncommon to find hollows at the base of the leaf where dust, dirt, and dead insects accumulate. One very curious plant of this sort is *Dischidia Rafflesiana*, in which the leaves have become quite like a pitcher, and have been compared to "natural flower-pots" intended to hold rain and leaf-mould.

Then there is the Bromelia or Pineapple family, which consists for the most part of plants which live on the branches of trees. In very many of these a small cup is formed in the middle of the rosette or tuft of leaves, and water collects in this central cup.

The water smells abominably, and contains the bodies of dead insects, and rubbish of all kinds. The remnants of these drowned insects are probably of use, because any valuable nitrogenous or other material may be absorbed with the water by the plant and help to nourish it, but in such a rough contrivance as this there is nothing comparable to the Side-saddle plant, Pitcher plant, and others.

The former, Sarracenia (or Side-saddle plant), is a common and rather widespread North American plant, which is especially abundant in Florida. It is cultivated in most botanical gardens, but can only be grown in greenhouses. The leaves are about six inches to a foot long, and are hollow, funnel-shaped tubes with a short, flat wing along one edge. They may be an inch or two in diameter at the top or wider end, where there is also a sort of half-open lid which keeps rain from getting into the inside of the leaf. The colour of these tube-like or vase-like leaves varies. It is often variegated with brown, red, and yellow, and is conspicuous enough even at a distance. Thus insects fly to these vases and alight on the little cap or lid, where they find honey and enjoy themselves. Other insects crawl up along the rim or wing of the vase, finding honey here and there along their road. Having got to the lid, the insect, being of an inquiring or inquisitive disposition, will look inside the tube and endeavour to find more honey therein.

It reaches the rim of the vase and finds that there is honey inside; it can easily crawl down, and fails to notice that the inside of the vase is lined with long stiff points which all point downwards. These points or hairs do not at all interfere with its passage down, and it proceeds to the honey which forms a smooth, slippery coating. Then, after greedily absorbing the honey, it tries to get out again. But that is quite a different matter. Each one of these points or hairs is facing it, and the whole inside is smooth and slippery. It struggles, slips, and falls into a pool of water which fills the lower part of the vase. That is what the plant has developed these pitchers for. The body of the insect after a time decays away, and only its empty shell remains. An extraordinary number of insects are caught by these Sarracenia vases. Sometimes in one which is only ten inches long, three or four inches will be full of the corpses of blackbeetles and other drowned insects, and it is said that birds occasionally visit these vases in order to pick them out. There is probably some sort of secretion in the water. "A centipede 1-2/3 inches long having fallen into a vase of *Sarracenia purpurea* in the night was found only half-immersed in the water. The upper half of the creature projected above the liquid, and made violent attempts to escape; but the lower part had not only become motionless, but had turned white from the effect of the surrounding liquid; it appeared to be macerated, and exhibited alterations which are not produced in so short a time in centipedes immersed in ordinary rainwater."

In some Sarracenias the vase is brought up into a sort of hood or dome with the entrance at one side and below. There are thin patches on this dome or cupola, and small insects, attracted by the light which comes through these bare places, remain dashing themselves against them or crawling over them just as flies do on a window-pane, until they become tired and fall down into the water below.

There is something horrible in the cold and careful way in which this plant arranges its baits for "confiding insects. The latter are fed with honey, even on the very border of the assassin's den, but after this farewell revel they generally slip upon the smooth edge, and are hurled, like lost souls, down into the abyss."

In another plant, the Pitcher Plant (*Nepenthes*, so called from the drug which produces the sleep of death), we find an even more beautifully arranged pitcher which acts in very much the same way. It is, however, only the end of a rather long leaf, or rather of its midrib, that is turned up to act as a pitcher. There are similar stiff hairs pointing downwards, and honey is plentifully secreted. But, in Nepenthes, there is also a distinct secretion which digests the bodies of the drowning insects. The ferment resembles the active principle of the gastric and pancreatic juices of the human body, and, as acids are also present, the insect's body becomes changed into nutritious juices which readily diffuse into the plant. Dr. Macfarlane found that when the pitchers were stimulated by being given insects, the liquid inside them could digest fibrin to jelly in from three-quarters to one hour's time. But certain insects have somehow managed to educate their larvæ to resist the gastric juices of Nepenthes.

Near Fort Dauphin, in Madagascar, I found great quantities of *Nepenthes madagascariensis*. Almost every pitcher was one-third to two-thirds full of corpses, but in some of them large, fat, white maggots, of a very unprepossessing appearance, were quite alive and apparently thriving. These must have been the larvæ of a blowfly similar to that which has been mentioned by others as inhabiting Sarracenia. At the same place a white spider was very often to be seen. Its web was spun across the mouth of a pitcher, and its body was quite invisible against the bleached remains inside.

It had suited its colour to the corpses within, in order that it might steal from the Nepenthes the due reward of all its ingenious contrivances!

A totally different arrangement is found in an inconspicuous and ugly little marsh and ditch plant called Utricularia or Bladderwort. It is very difficult to see, for unless it happens to be in flower it is entirely submerged in the water. The flowers, which are purple, are conspicuous and easily seen even at a distance. On these submerged leaves there are hundreds of small bladders. They are about the size of a pea, and are most ingeniously contrived to catch small water-animalcula. The general idea of the bladderwort is exactly that of the eel-pots so common in some parts of the Thames. There is a small flap which acts as a trapdoor. Small creatures probably take refuge in the bladders when pursued by the larger water-fleas, etc., for it must seem to them to be a safe and secure retreat.

But once within the door, they are imprisoned and cannot find their way out again. They perish inside and their bodies are digested by the plant; on the inside of the bladder there are gland hairs which also secrete a digestive fluid.

The bladderwort is dangerous to fish, for the little fry, when quite small, run their heads and gills into the bladders and are suffocated.

There are a great many kinds of Utricularia, and they occur in most of the great floral regions.

One of them has chosen a very extraordinary and curious situation. It lives inside the little cups of water which, as we have already mentioned, are formed by the leaves of some Bromeliads. The insects in the water which ought to nourish the Bromeliad (*Tillandsia*) are really used by the Utricularia. Other Utricularias live in damp earth, moss, etc.

It is not only by traps and pitfalls that plants catch insects: many have specially modified hairs which are quite efficient insect-catchers.

Hairs are used by plants for many different purposes, and it is rather interesting to see how quite a simple organ like a hair can be altered. The stinging hair of the nettle has already been mentioned; many grasses possess minute, rough, flinty hairs, which probably prevent snails from eating them. That also is probably the reason of the strong, rough, coarse hairs which cover the Borage and the Comfrey.

Then on the Chickweed and the Bird's-eye Speedwell there are lines of rather long, flexible hairs which at first sight appear to be of no use at all. But if you take either of these plants, and, holding it upright, place a large drop of water on the leaves, you will see that these hairs are intended to carry the water down the stem. The water runs along them. It is a very pretty little experiment, especially if done in artificial light, so that these hairs are, like the root hairs, intended to absorb or suck up water as it passes over them. Then the Edelweiss and the Lammie's Lug (*Stachys lanata*) are entirely covered with white cotton-woolly hairs: these are intended to keep the water in the plant, and do so as effectually as a rough woollen coat will keep out rain and mist. Silky hairs, downy hairs, and others are found wrapping up the tiny baby leaves in the bud: they probably keep them warm, and perplex and ward off objectionable insects.

But, perhaps, the sticky or glutinous hairs are the most wonderful of all. They are found on many plants, such as *Salvia glutinosa*, Plumbago, and Catchfly. One can see insects stuck on them and vainly struggling to be free, and the hairs undoubtedly prevent green-fly and other such pests from interfering with the honey of the flower. In some of these cases it has been shown that the body of the insect is actually used as food, but that is more obvious with two interesting plants which specially devote themselves to the capture of insect prey. One of these is very often kept in the Boer farmhouses near Tulbagh, in South Africa, simply to attract the flies, which are a perfect pest in those dry valleys. Another Drosophyllum, the Fly-Catcher, grows on sandy and rocky ground in Portugal and Morocco. This is also used by the peasants near Oporto as a convenient fly-paper.

In both of these plants large drops of a sticky, glistening liquid are secreted by the hairs which cover the leaves. Any small insect alighting on the latter is sure to get covered by the liquid, and in trying to get away will become hopelessly involved in it. It is probably soon suffocated, for the gummy matter will choke the small air-holes by which it breathes. Both these plants are said to secrete both an acid and a digestive secretion.

But we have two plants which are even more interesting in this country.

Walking over the rough marshy pastures or moors of Scotland one is sure to notice, generally on wet peaty and barren soil, a little rosette of bright, yellow-green, glistening leaves. If it is the right season there will be a handsome purple flower whose stalk springs from them. This is the Butterwort (*Pinguicula*), and it is not a bad name, for the leaves remind one of butter. The whole upper surface of the leaves is covered with tiny glands secreting a sticky, glistening matter. It is said that there will be as many as fifty thousand of these glands on a square inch of the upper surface.

Now in such places every one knows that there are quantities of midges, and also that these insects are always exceedingly thirsty. They prefer blood, it is true, but when they see these bright yellowish leaves they naturally go to them. When, however, the midge touches the leaf, the sticky liquid clings to its wings and legs, and it cannot escape.

So far this does not differ from the Fly Catchers mentioned above, but another very curious action then begins. If the midge or fly is near the margin of the leaf, the edge of the latter begins to curl or roll inwards over it. It does so very slowly, and may not finish rolling over the insect for some hours. Whilst this is going on acids and "gastric juice," or ferments which act in the same way, are being poured over the body of the midge, which is finally completely digested. Next day, having finished the midge, the leaf majestically unrolls itself again and waits for another.

The juice contains rennet, and is used by the Lapps in making a horrible delicacy called Tätmiölk. It has also been used by the Swiss shepherds for at least two hundred years, to cure sores on cows' udders.

The other British plant is the Sundew (*Drosera*). Every one who has been on peat-mosses and moors probably knows its little reddish rosettes of small rounded or spoon-shaped leaves lying on bare peat or wet mossy ground.

Each leaf seems to be covered by hundreds of glittering little dewdrops (whence the name).

The hairs or tentacles which cover the leaf secrete this glistening, sticky fluid. There must be about two hundred of them on a single leaf.

An insect flying about near the Sundew is sure to be attracted by the conspicuous glittering, reddish leaves, and probably alights upon it. Then it finds itself caught and begins to struggle, but this simply brings it against more tentacles.

Now happens the most wonderful part of the whole performance. All the neighbouring tentacles, although they have not been touched, bend over towards the struggling insect and pin it down in the middle of the leaf. They do not bend over very quickly. In two or three minutes they will bend over towards it through an angle of forty-five degrees, and it takes them ten minutes to bend over ninety degrees.

There is something rather horrible in the sight of a large insect struggling with these slow, remorseless, well-aimed tentacles; most people free the insect unless, at least, it happens to be a midge. The point which is so difficult to understand is to know how those untouched tentacles know that the insect is there and exactly where it is. There is no doubt that they do know, for they behave exactly as if they were the arms of a spider.

If you put two insects on either side of the middle of the leaf, half the tentacles will pin down one and the other half will deal with the other insect.

At the same time acids and ferments are poured out which digest the insect. It takes about two days for a leaf to finish off an insect, and then the tentacles again unclose.

Moreover it is difficult to deceive those tentacles. They will bend in for the tiniest piece of useful substance; for instance, a length of one-seventy-fifth of an inch of woman's hair will make them secrete digestive fluid. One millionth part of a pound of ammonium phosphate will also produce secretion. But a shower of heavy rain, grains of sand, or other useless material, will not cause any secretion, and even if they do bend in a little, they soon discover their mistake and stand out again. If you try the same experiment under a bell-glass from which the oxygen has been withdrawn by an air-pump, nothing happens; or if you chloroform the Sundew it will pay no attention to small pieces of meat until it recovers from the effects of the chloroform.

When these Droseras are taken to a greenhouse and experiments are made on them, they run into very great danger. They are almost certain to die of overfeeding or indigestion. It is impossible to keep people from giving them too much to eat.

This wonderful little plant shows quite distinctly that there must be some way of sending messages in its leaves. Somehow the message travels from the tentacle which the fly has touched, down the stalk into the leaf, and up into the other tentacles, and tells them that there is something worth stooping for.

No one has explained this, and probably no one will ever do so.

The last, and in some ways the most interesting, of all these carnivorous plants is Venus' Fly-trap (*Dionæa muscipula*), which grows in North America from Rhode Island to Florida.

It is a quite small herb with a small circle of leaves which lie flat on the ground. Each leaf ends in a nearly circular piece which is divided by a very marked midrib. The two semicircular halves have a series of teeth along their edges; these margin teeth are stiff and a little bent upwards. In the centre of each half there are three small hairs. On looking closely at these hairs one finds that each has a joint near the base; all over the centre of these leaf halves there are scattered glands which secrete ferments intended to digest any animal matter.

The really interesting point is connected with these central jointed or trigger hairs; they are extremely sensitive. But when they are touched it is not they themselves that are affected, but the entire circular end of the leaf!

Suppose an insect wanders on to the leaf and reaches one of these semicircular halves, nothing happens until it touches one of these hairs, but then *both* halves suddenly close together, exactly like an ordinary rat-trap! The teeth on the edges of the halves interlock like the teeth of a trap, and the insect is caught and imprisoned.

Its body is slowly digested away and goes to nourish the plant. The use of the joint in the sensitive hairs can be easily perceived, for when the two halves shut up together, the hairs fold down exactly like the funnel of a river steamboat when it passes under a bridge.

The closing of the two halves, which has been well compared to shutting up a half-open book, is very quick, as it does not take more than ten to thirty seconds. There is an abundant flow of "gastric juice," but the leaf takes a long time to digest its food. It may require three weeks to finish one insect. Moreover, if overfed, it may turn a bilious or dyspeptic yellow colour, and wither or even die. It only shuts for a short time if a grain of sand touches the sensitive hair, and, like Drosera, is not deceived in its food.

The Dionæa, Drosera, the Sensitive Plant, Mimulus, Barberry, and others, all show us clearly that plants somehow or other act as if they were conscious of what they ought to do. In fact, in all these cases, it is scarcely possible to help believing in some sort of rudimentary nervous system. At any rate Wordsworth comes near this belief, for he has written:—

"It is my faith, that every flower that blows
Enjoys the air it breathes."

CHAPTER XXVIII. MOSSES AND MOORS.

PEAT-MOSSES AND THEIR BIRDS—MOORLANDS—COTTON-GRASS—SCOTCH WHISKY—GROWTH OF PEAT-MOSS—A VEGETABLE PUMP—LOW-LYING AND MOORLAND MOSSES—ERUPTIONS AND FLOODS OF PEAT—COLONIZING BY HEATHER AND SCOTCH FIR—PEAT-MOSSES AS MUSEUMS—REMAINS OF CHILDREN AND TROOPERS—IRISH ELK—STORY OF THE PLANTS IN DENMARK—RHODODENDRONS AND PEAT—USES OF PEAT—RECLAIMING THE MOSSES NEAR GLASGOW.

IN Great Britain in this present year one finds exceedingly few places where the influence of man cannot be traced. Over most of the country, indeed, it is impossible to discover a single acre of land where Nature has been allowed to go on working at her own sweet will without interference or restraint.

But near Stirling, between the Lake of Monteith and the sea, there is a wide, desolate valley which is probably in exactly the same condition as it was when the Roman legions halted to reconnoitre before Agricola passed onwards to Perth and Aberdeen.

Indeed, this great peat-moss has been probably in very much the same condition for some 200,000 years, which is a nice round number to represent the ages that have passed since the Great Ice Age.

Now, as then, it is inexpressibly dreary and desolate; everywhere saturated with water, and only to be traversed in dry seasons and with much agility. Even with the greatest care the pedestrian may sink to the waist in a hole of black, slimy, peaty water. Moss, Heather clumps, Sedges, Rushes, and occasionally Cotton-grass, almost at one dead level, stretch right across from the one side of the huge valley to the other.

Even grouse are not common. In summer great numbers of gulls lay their eggs upon the moss. This also is one of the few places in Britain where great flocks of wild geese can be heard and seen, but only at a distance.

It is almost impossible to get near them, for the upright neck of the sentinel cannot be seen by the stalker as he wriggles towards the flock on his face, until long after the stalker himself has been plainly visible to the bird.

Of all useless stretches of barren waste, such a moss as this seems one of the worst. It would, of course, be possible to reclaim it; probably, fertile fields and rich meadows *could* be formed over the whole valley, but it would not pay nowadays. There is so much good land available in Canada, the United States, and Australia, that this great stretch of our native country will probably remain as useless as it was in Agricola's days.

In the Scottish Lowlands and Highlands the moorlands are almost as desolate. At a height of 1500 to 1600 feet in Southern Scotland there is nothing to be seen but the undulating lines of hills, all dark purple with heather or with the peculiar scorched reddish green of Deer's Hair and dried sedges.

Perhaps on the nearer hills small streams may have cut a whole series of intersecting ravines in the black peat. They may be six to ten feet deep, and here and there the bleached white stones which underlie them are exposed. Now and then the "kuk-kuk-kuk" of an irate cock grouse, and much too frequently the melancholy squawking of the curlew, irritates the pedestrian as he stumbles over clumps of heather, plunges in and out of the mossy holes, or circumvents impossible peat-haggs.

This is Draba Alpina from Cape Tscheljuskin, and it is drawn the natural size. The stunted, closely set leaves show the inclement character of the climate.

It is indeed a remarkable fact that though these islands support 44,000,000 of inhabitants, including at least 1,000,000 paupers and unemployed, one-seventh of Ireland and many square miles in Scotland are still useless peat-bogs!

The Bog of Allen alone covers 238,500 acres, and the peat is twenty-five feet deep.

In some few places the peat is still used for fuel, and there is a theory to the effect that peat reek is necessary for the best kinds of Scotch whisky, but neither grouse nor black-faced sheep, which live on the young shoots of the heather, employ in at all a satisfactory way these great stretches of land.

Many attempts have been made to spin the silky threads of the Cotton-grass which grows abundantly on the Scotch lowlands. It is neither a grass, nor does it supply cotton, but is called Eriophorum. It is perhaps the one really beautiful plant to be found on them, for its waving heads of fine silky-white hairs are exceedingly pretty.

The heather itself gives a splendid red and purple shade, which in summer and autumn is always changing colour, but it is monotonous. Neither the little Bog Asphodel with its yellowish flowers, nor red Drosera, or butter-coloured Butterwort, are particularly beautiful.

After seeing such a country one understands something of the Cameronian Covenanters who held their conventicles and took refuge therein.

The manner in which these mosses and moors have developed is most interesting, and yet difficult to explain.

There are two kinds of peat-mosses, which, although there are many intermediate types, may be kept apart.

The first, like the one near Stirling, Lochar and Solway Moss, near Dumfries, and Linwood, near Glasgow, have been formed in low-lying flat estuarine marshes.

If one refers back to page, it will be seen how reeds and rushes and marsh plants may gradually fill up river backwaters. Eventually a saturated, marshy meadow is produced.

Then comes the chance of that wonderful moss the peat-moss, or Sphagnum. It is scarcely possible to appreciate its structure without the help of a microscope and a good deal of trouble in the way of imagination.

It is in a small way a sort of vegetable pump which raises water a few inches or so. Stem and leaves and branches possess little cistern cells, which act both as capillary tubes raising the water and also retain it. The stems are upright and develop many branches, so that they become a close-ranked or serried carpet of upright moss-stems squeezed together, which floats on the surface of the water. Each moss-stem is growing upwards and dying off below. In consequence, the bottom gets filled up by dead mossy pieces, which accumulate there, while the live moss-carpet remains floating on the surface of the loathly, black, peaty water.

In many peat-mosses the water gets entirely filled up, but that does not stop the formation of the peat-moss. It is now resting on the water-saturated remains of its forefathers, and if water is abundantly supplied it goes on developing.

Thus in these lowland or estuarine peat-mosses the moss eventually occupies the water, and goes on growing. After this it develops like the moorland mosses which cover most of the Lowlands and Highlands of Scotland. They cover the hills, and it looks exactly as if some giant had plastered all those hills with a layer of six to ten feet of black peat from 1250 feet upwards.

The soil would at first be covered by a saturated moss-carpet of *Sphagnum* and other mosses. Rainwater falling upon it was all retained, and very little could get away, for the Sphagnum carpet is just like a huge sponge soaking up and retaining the water.

But it sometimes happens in these great upland mosses that there are enormous falls of rain which continue for days. Then the water collects *under* the living moss-carpet and over the dead peat. It may be gathered together in such quantities that the carpet of living peat above it bursts, and a deluge of peaty water overflows the surrounding country, destroying and spoiling everything that it encounters.

The worst of these inundations of black mud that has happened in recent years was in December, 1896, near Rathmore, where 200 acres of bog burst and a horrible river of mud overflowed the country for ten miles. Nine people perished, and enormous destruction was caused.

There have been many other cases. In 1824 Crowhill Bog, near Keighley, burst; and in 1745, in Lancashire, a space a mile long and half a mile broad was covered by peaty mud. There was also a case in 1697, where forty acres of bog at Charleville burst in the same way.

Attempts have often been made to calculate the rate of growth of such peat-mosses. A great many of them began to develop on the mud left by the ice-sheet when the glaciers retreated at the end of the Ice Age. Those mosses are therefore probably 200,000 years old. Some of our Scotch mosses are twenty to twenty-five feet in depth, which gives a foot in 10,000 years. By calculation of the weight of the peat formed, Aigner made out that a certain moss was 20,600 years old, and was growing at the rate of two inches in a century.

But in Denmark ten feet has been formed in 250 to 300 years, and in Switzerland three to four feet of peat-moss has been formed in twenty-four years.

This shows quite distinctly that there is no regular rate of growth, and indeed it is obvious that much must depend on the climate, on the rainfall, on the drainage, and other circumstances.

Sooner or later, however, a limit comes to the growth of the moss. The surface then becomes gently curved: it is highest in the centre, and slopes very gently down in every direction to the edges.

What happens next? The first sign is that the surface begins to dry up, and Heather, with grey Cladonia lichens, begins to grow on the projecting tufts and tussocks.

Occasionally, if gulls build their nests on such drying-up mosses, patches of bright green grass appear wherever the gulls are in the habit of resting. That is due to the lime in their guano.

But under quite natural conditions a much more important and interesting change begins.

Here and there scattered over the moss, miserable little seedling Birches and Scotch Firs begin to struggle for life. Of course, if there are hares and rabbits, or if sheep and cattle are allowed to graze upon the moss, those firs have no chance whatever. They are eaten down to the ground.

The Irish elk is the result of the day's sport of these prehistoric Britons, who lived in houses built on piles actually in the water, or in peat mosses. Their only boats were rough dug-out canoes.

But if allowed to go on growing they would no doubt cover the whole moss with a wood of Birch and Scotch Fir. In time that wood would by its roots and its formation of fine leaf-mould so radically alter the ground that a forest of Oaks might be possible.

It is in fact quite likely that most of our Highland and Scotch hills were at one time covered by fine forests of Scotch Fir, of which the *Silva Caledonica* spoken of by Tacitus was an example.

There is, moreover, evidence to show that this was the case. There is one strange peculiarity of peat which renders it a most useful substance to antiquarians.

Anything lost in a peat-moss does not decay away, but remains in a blackened but still recognizable condition for hundreds of years. Not long ago a basket containing the bones of a child was found in a Scotch peat-moss. There is also a story that an English trooper of the fourteenth or fifteenth century, and his horse, were discovered in Lochar Moss, near Dumfries. The man's features were traceable at first, but fell into powder when exposed to the air; but the weapons, stirrups, etc., were all perfectly preserved. Bones of the extinct Irish elk have often been found. Not merely so, but the piles of lake dwellings and the rough dug-out canoes which were used by the early inhabitants of Britain have been discovered in a great many places. Coins of Roman, medieval, and modern times have been unearthed, and indeed there is no doubt that if Britain is still inhabited two thousand years hence, boots, sardine tins, brass cartridges, clay pipes, and other characteristic products of our own days, will be disentombed from the peat by enthusiastic antiquarians, and displayed in museums to admiring crowds of our descendants.

The reason is quite simple: in peat neither those bacteria which cause ordinary decomposition, nor worms of any kind, are able to exist, so that the material does not decay but accumulates, though it may be blackened by peat, water, and humic acid. It is for this reason that a peat-moss is such a bad or rather an impossible soil. Neither roots nor bacteria can thrive in saturated peat; therefore the flora of a peat-moss is generally confined to the upper surface, where air and bacteria can reach the roots. Peat-mosses are also the home of insectivorous plants, which get their nitrogenous food from the insects which they catch.

In consequence of this preserving effect of peat, it is possible to trace the entire history of a peat-moss from the very beginning. Remains of the Dwarf Willow or Polar Birch have been found in England, showing that those now Arctic plants were then flourishing in Norfolk. These are generally in the lowest layers of peat-mosses. Next follow remains of the Birch and Aspen, which would be growing, as they do in places to-day, on mossy soil where the peat was still thin. Higher up in the peat one finds remains of Scotch Fir, showing that at that time regular forests of Scotch Fir existed, e.g. in Sutherlandshire and on Lochar Moss, where they do not grow at present.

Some hold that the goats, black cattle, and ponies which have been kept since the Roman occupation at any rate, are responsible for the destruction of these forests. Others hold that they were killed by a change of climate. But they certainly existed.

Trunks of Scotch Fir have even been found in peat at 2400 feet in Yorkshire, and at heights in Scotland which are above all the present plantations. About this time it seems that the newer Stone Age men must have been in Switzerland and Denmark, for their remains and characteristic weapons occur in those countries at the same level in the mosses as the Scotch Fir.

Still higher in the peat comes the Bog Oak. With it are in Denmark remains of the Bronze, Iron, and Roman times.

In Denmark the uppermost layers of the peat contain remains of Beech trees. As this last tree only entered the country in the historic period, it is not found except in the highest layers of all.

Unfortunately we have not yet obtained in our own country the same evidence from the peat-bogs as to the history of the flora of Britain. It is at least probable that it was on very much the same lines.

Would it be possible to again cover our peat-mosses and moorlands with forests of Conifers, Pines, Larches, and Spruces? There can scarcely be any doubt about it: it would be possible, and according to the best authorities it would even pay to change all land which is not yielding more than 7s. 6d. an acre into forests of Pines.

One of the curious facts about peat is that though a peat-moss is one of the worst natural soils, yet broken-up and dried peat is excellent for Rhododendrons, for Orchids in stoves and greenhouses, and a great many other plants.

Peat consists of very much the same substances as those that go to form leaf-mould. But the presence of humic and other acids, and the saturation with water and consequently the absence of worms, bacteria, and also of air, make it impossible for plants to grow in a peat-moss.

Peat-moss due specially to the Cotton-grass rather than the Sphagnum moss is imported in great quantity from Holland, for use as litter for horses. We have in this country plenty of peat quite good for this purpose, but labour is too expensive for our home-grown peat to compete with the produce of Dutch moors.

But that is by no means all the uses to which peat can be put. It is interesting to mention a few of them.

1. Peat is used as fuel.

2. Growing Orchids, etc.

3. Litter for poultry, cattle, and horses.

4. Food for cattle, etc., is made by rubbing the peat into small pieces and saturating with molasses.

5. Paper and a kind of felt can be made of peat.

6. Rugs and carpets can be made of peat-fibre.

7. String and twine.

8. Rough sacks and mats can be made of peat-fibre.

Unfortunately, though all these things can be produced out of peat-fibre, it has never paid to manufacture them, and there are very few of the British peat-mosses nowadays where peat is even cut for fuel.

It seems much more likely that the end of these peat-mosses will be to become either agricultural land or forest.

Near Glasgow a large area of a useless peat-moss has been reclaimed and made to yield excellent crops, by using the refuse of the city. The disposal of such refuse used to be a most troublesome and expensive process, but now it is turned to good effect.

It was suggested a few years ago that peat, which is not worth conveyance, should be burnt on the spot, and the energy transmitted by wires.

That would be quite impossible, in at least four years out of five, over most of Scotland.

CHAPTER XXIX. NAMES AND SUPERSTITIONS.

GIVING NAMES THE FIRST AMUSEMENT—CURIOUS AND ODD NAMES—A SPITEFUL NATURALIST—THE MELANCHOLY BARTZIA—COMMON NAMES—BRITISH ORCHIDS—DANCING GIRLS AND COLUMBINES—SUSANS—BIBLICAL NAMES—ALMOND, APPLE, LOCUST—SPIKENARD—TARES—EFFECTS OF DARNEL—DAFFODIL—ACANTHUS LEAF—GHOST-DISTURBING BRANCHES—ELDER OR BOUR TREE—ITS POWERS AND MEDICINAL ADVANTAGE—DANEWORT—MANDRAKE—HOW TO PULL IT UP—THE INSANE ROOT—ITS PROPERTIES—PLANTS WHICH MAKE BONES PINK—THE BETEL NUT—HENNA—EGYPTIAN AND PERSIAN USES—CASTOR OIL—LEEKS, ONIONS, AND GARLIC—ANCIENT USE OF THEM.

MAN has always taken a certain pleasure in giving names to both plants and animals. It was, of course, a necessity to do this, but it is probable that people enjoyed the process as they do now.

At the present moment there must be at least 200,000 plants named and described by botanists. So that the number of ecstatic moments enjoyed by humanity has been undoubtedly increased.

The Egyptians, the Babylonians, and the Arabs named a great many plants, but for the most part those names are quite lost. Most of the knowledge of the Egyptians and Babylonians remained a close secret confined to their priestly colleges or universities, and has entirely perished.

For centuries those fragments of the knowledge of Greece and Egypt which were preserved seem to have been translated and taught in Latin. Long after the Roman Empire had passed away, all knowledge, including that of medicine, of botany, and of law, was imparted in Latin, which indeed was supposed to be learnt by every educated person almost until the present century.

Even now descriptions of new plants have to be given in Latin, and the name must have a classical appearance. Of course, nowadays, it would be much more convenient and much more generally useful if every person learnt English, German, French, and Japanese, but in this case of naming plants, the Holy Roman Empire still exercises its sway over the whole world.

Very often the names given to plants are of the most extraordinary character. The Latin is curious and the Greek remarkable, yet sometimes they are both pleasant to the ear and have a pretty and poetical meaning.

Poggeophyton, on the other hand, for example, means the plant discovered by Dr. Pogge, a German botanical explorer. *Wormskioldia, Zahlbrucknera, Krascheninikowia, Acanthosicyos, Chickrassia, Orychophragmus, Warczewiczia, Lychnophoriopsis, Krombholtzia, Pseudorhachicallis, Sczegleewia, Zschokkia,* are all names that sound harsh and look odd to us. Yet most of them are just called after those who discovered them, or their friends. In many of the smaller microscopic plants the names are really much longer than the plants themselves. Thus *Pseudocerataulus Kinkeri* is a diatom which cannot possibly be seen without the use of a microscope.

Names are and were given in the most extraordinary way. Not merely great botanists, but Themistocles, Aristides, Aristobulus, Virgil, and even Gyas and Clianthes, have plants named after them.

Yet that is not inexcusable, if people had not sufficient inventive power to do better. There was a naturalist who quarrelled with the great French scientist Buffon. Therefore he baptized as *Buffonia* a group of ugly, unimportant little plants which had an unpleasant smell. In other cases people have named plants after their sweethearts or friends.

A British plant called Bartzia has a rather melancholy, desolate appearance. It was named when the author had just received the news of the death of his friend Dr. Bartsch.

One of the most usual complaints which one hears from those who are beginning to study flowers is that the Latin names are so difficult and hard to remember. But they are not really more difficult than the common popular names, and especially those of foreign plants. Cheirostemon, for instance, which means stamens like a hand, is much easier to speak and to remember than *Macpalxochitlquahuitl,* which is its soft, meandering, Spanish-American name. Asperula (little rough one) is quite as good as Squinancywort, which means a herb good for quinsy (it is moreover of no good in quinsy). Perhaps, however, Woodruff (which is really "wood rowel," from the resemblance of the leaves to an old-fashioned spur), or

Waldmeister (master of the woods), are as good names as Asperula. Then Erigeron, which means "soon growing old," is an excellent description of the faded appearance of this little weed, for which the popular name is Fleabane (it has no effect upon these creatures whatsoever).

How popular names came to be associated with particular flowers is generally quite unknown. A fair number are called from the diseases which they are supposed to cure. Lungwort, however, was so called because the lichen Pulmonaria has a resemblance to lungs. Then in course of time people began to suppose it was a cure for diseases of the lungs, which it is not.

The British Orchids are called Bee, Spider, Fly, and Hanging-man Orchids, because of a fancied resemblance to their namesakes. Dancing-girls (*Mantisia*) was so called from a certain resemblance of the flower to a columbine. The true Columbine (*Aquilegia*) was so called because of a resemblance which some one saw to a circle of little doves with wings seated on a circular well.

The greatest objection to popular names, however, lies in their being so indefinite. Entirely different plants are known by the same name, and also in different parts of the country totally different names are given to the same plant. All such difficulties disappear if one takes the trouble to learn the Latin names.

These also are often quite pretty. Luzula, Veronica (with its pretty legend), Mimulus (the little monkey), Circæa (Enchanter's Nightshade), Senecio (*the old man*, from its woolly head of fruits), Nymphea, Naias, Carlina (*the old witch*), and so on, are quite as pretty and as nice as Mugwort, Devil-in-a-bush, Hairy Rock Cress, and the rest. One curious result of the use of popular names is seen in the Biblical names of plants. The Rose of Sharon seems most probably to have been *Narcissus Tarzetta*, and not a rose at all. As regards the lilies of the Field, Mr. Ridley has the following remarks. The Hebrew word *Shushan* was a generic name given to a mixture of flowers, exactly as we now talk of ferns, herbs, or grass. The Sermon on the Mount was preached near the plain of Gennesaret, and there flourish the Anemone (*Anemone coronaria*), *Ranunculus asiaticus*, and *Adonis aestivalis* and *flammea*, which are exactly of the same colour and follow each other in close succession. This word *Shushan* is the original of the Christian name Susannah or Susan. The Arabic name for *Anemone coronaria* is Susan.

The Almond of the Bible is the common almond which is wild in Syria and Palestine. "Aaron's rod that budded was a branch of an almond tree; the bowls of the Golden Candlestick were designed from the almond blossom. Even at the present time English workmen call the glass drops for ornamenting candlesticks almonds." The Apple of the Bible was more probably an Apricot. The husks of the prodigal son were probably the Locust-beans, sometimes called St. John's bread, but it is quite probable that the "locusts" eaten by St. John were the insects. At any rate, locusts are regularly eaten in the East. The Locust Tree (*Ceratonia siliqua*), or Algaroba or Carob, has large, dark-purple pods; there is a pulpy material between the seeds which forms a valuable cattle food. The seeds are said to have been the original "carat" weight of jewellers.

The Spikenard (*Nardostachys jatamansi*) belongs to the natural order *Valerianaceæ*. It is a wild plant of Bhutan found near Rangasnati, in India, and in ancient times it was transported on camels by the regular caravan route to Syria, Greece, and Rome. It was then worth about £3. 10s. per lb. The essence is obtained from the roots, but one hundred pounds of roots will furnish only half a pound of essence. Now it has but little value.

The Tares sown amongst the wheat were probably the seed of the Darnel. When growing, this grass is very like wheat, and it would be quite possible to mistake one for the other until the flowers and fruit are formed. Darnel is one of the very few poisonous grasses. It is said that the poison is produced by a fungus which is found in the grain. When darnel seed is ground up with wheat the bread becomes dangerous, for the poison produces severe headache, vertigo, and giddiness. Other authorities say that it causes in man and rabbits eruptions, fits of trembling, and confusion of sight. It seems not to affect horned cattle, swine, and ducks.

As regards those plants which were specially beloved and venerated by the Greeks, there is not very much to say. Moly seems to have been *Allium moly*, one of the onion or garlic family. It is not very remarkable in any way. Amaranth was apparently the garden Love-lies-bleeding, called in France Queue-de-Renard and Discipline-de-Religieuse. The Asphodel which covered the Elysian fields seems to be *Asphodelus ramosus*. This grows in quantities in Apulia, and is said to afford good nourishment for sheep.

The Myrtle, with which the Athenian magistrates and victors in the Olympic games were crowned, is not really a European plant, though it has a wide range from Asia Minor to Afghanistan. It was sacred to Venus, and had some importance as a medicinal plant and for perfumes. It was even used in cookery and for making myrtle wine, which last is said to be still prepared in some parts of Tuscany.

"Narcissus, son of the river Cephisus and of Liriope, daughter of the Ocean, was a young man of great beauty who scorned all the Nymphs of the country, and made to die of languor Echo, because he would not respond to her passion. But one day returning from the chase weary and fatigued, he stopped at the side of a fountain to refresh himself. There having seen his own face in the water, he was so smitten with it and so greatly loved himself that he died of grief. The Gods, touched by his death, changed him into a Daffodil, according to the fable."

Such is the account in M. l'Abbé Ladvocat's *Dictionnaire Historique-Portatif*, Paris, 1760. Daffodil means appearing early in the year. The number of races, varieties, and forms of Daffodil, Jonquil, etc., has become innumerable; yet it is doubtful if any are quite so graceful and absolutely charming as the *Narcissus poeticus*, supposed to be the original of the above legend.

The Acanthus leaf which was so much used in sculpture seems to have been that of *Acanthus spinosus*. It can still be traced in modern carving, though, of course, it is very much altered and in a rather degenerate form.

It is often very difficult to say why certain plants have received so much attention and veneration in ancient times. In some cases it is clearly because they are poisonous, and therefore become dreadful and awe-inspiring. Why, however, should a twig of Rowan (*Pyrus Aucuparia*) be so often placed above the door of a Highland cottage? In some way it was supposed to keep off evil spirits, but there is no special reason why it should have been chosen.

The "Bour Tree" or Elder (*Sambucus*) has been the centre of a whole series of extraordinary and remarkable superstitions. Of the Ellhorn (Low Saxon), or *Sambucus nigra,* Arnkiel gives the following account: "Our forefathers also held the Ellhorn holy, wherefore whosoever need to hew it down must first make his request, 'Lady Ellhorn, give me some of thy wood, and I will give thee some of mine when it grows in the forest'—the which, with bended knees, bare head, and folded arms, was ordinarily done."

The flowers are an eye-wash and cosmetic, or they may be taken as tea or used as a fomentation. The berries are used for "elderberry wine."

A certain cure for rheumatism is to carry about a small piece of elder cut after the fashion of a rude cross.

Evelyn speaking of it says: "If the medicinal properties of the leaves, bark, berries, etc., were thoroughly known, I cannot tell what our countrymen could ail for which he might not fetch a remedy from every hedge, either for sickness or wound."

The other species (*Sambucus ebulus,* or Danewort) has had its name explained as follows by Sir J. E. Smith: "Our ancestors evinced a just hatred of their brutal enemies the Danes in supposing the nauseous, fetid, and noxious plant before us to have sprung from their blood."

Of all these, however, the Mandrake (*Mandragora*) is connected with the most extraordinary and remarkable superstitions. The plant is distinctly poisonous, and has peculiar divided roots which sometimes have a very rough resemblance to the human body. It was supposed to be alive, and to utter the most piercing shrieks when it was pulled out of the ground. In those accounts, which are based on that given by Josephus, it is the *person* who pulls out the root, and not the plant, that shrieks, subsequently rolls on the ground, and finally dies in torments. Therefore, if you wish to pull up a mandrake, the correct course to pursue is as follows: Tie a dog to the plant by its tail, and then whip the dog. It will pull up the mandrake, and then die in frightful agony!

This is the "insane root" of Macbeth, but its various uses, real or pretended, are too numerous to explain in detail.

Thus it was used for the following purposes: as a poison, an emetic, a narcotic like chloroform, in love-philtres and love-charms, as well as to dispel demons, who cannot bear its smell or its presence. There are many of these relics of medieval times which are difficult to explain or to find a reason for.

Why, for instance, should old women always carry a sprig of Southernwood to the kirk in their Bibles? The leaves are, however, said to be disagreeable to insects. The Lavender stalks usually placed in linen both keep away insects and have a pleasant old-world scent.

A great many of the properties possessed by plants are of the most extraordinary and unsuspected nature. The roots of the Madder (*Rubia tinctorum*), for instance, when they are eaten by swine or other animals, change the colour of their bones, which become pink. This curious property has actually been made useful, for physiologists have employed madder in the study of the growth and development of bone.

In India and other eastern countries one is often shocked and surprised to find an apparently quite healthy native expectorating blood in a most lavish manner.

But the native is only chewing Betel nuts, which have the power of turning the saliva red. The fruit is that of *Areca Catechu*, a fine palm which is cultivated, for this purpose only, in many parts of India and the East. The nuts are cut in pieces and rolled up with a little lime in leaves of the Betel pepper. It is said to turn the teeth red and sometimes to produce intoxication; at any rate, people become slaves to this disgusting habit, and they do not seem to be at all injured by indulgence in it.

Another extraordinary plant is Henna (*Lawsonia inermis*). The Egyptian mummies are found to have the soles of the feet, as well as the palms and finger-and toe-nails, dyed a reddish-orange colour by the use of henna. But the practice is continued to-day in most parts of the East, and no odalisque's toilet would be considered complete without the use of henna. It is even said that men dye their beards with it.

The white horse used in processions by the Shah of Persia has its legs, tail, and body dyed with henna.

The powdered leaves are used: they are made into a paste with hot water, and then spread upon the place. It is grown in Syria, Egypt, Algeria, China, Morocco, Nubia, Guinea, and the East Indies.

In China women dye their eyebrows with an extract of the petals of *Hibiscus Rosa-sinensis*.

One of the first plants to be utilized by man was the Castor-oil (*Ricinus communis*). It was used by the Indians from time immemorial; it is mentioned by Herodotus (under the name Kiki); seeds have been found in mummy-cases, showing the careful preparations which were made for the dead when starting on their travels in the other world!

It is one of the very commonest plants in the tropics and in sub-tropical or warm, temperate countries. It is rather handsome, and has large reddish-green leaves and handsome spikes of flowers. It is said to be sometimes twelve feet high, but is usually only six or seven feet. The seeds are mottled or marbled, and have a distinct resemblance to a beetle when seen from above. It has been suggested that this protects them from birds, or enables the latter to recognize the seed, which is strongly medicinal. That, however, is at least doubtful, and certainly pigeons are exceedingly fond of the seeds and eat them in quantity. The oil is used for lighting, in making soap, and also in painting.

Another characteristic Egyptian plant was the Leek, which with the onion and garlic seems to have been one of the very first to be brought into cultivation. Herodotus says that on the Great Pyramid there was an inscription saying that 1600 talents had been paid for onions, radishes, and garlic used by the workmen during its erection.

The Jewish priests were forbidden to eat garlic, which (with cucumber) formed the dishes most regretted by the Israelites during their wanderings in the wilderness. The Shallot comes from Ascalon, where it will be remembered Richard the First defeated Saladin the Sultan, and where also Sir Sidney Smith defeated the Emperor Napoleon and made him miss his destiny. It was not brought to this country till 1548. Probably, therefore, Tennyson's Lady of Shalott lived somewhere else. Onions and leeks are of course popular in this country, and especially in Wales, where the latter has been the badge of the Welsh since they gained a victory over the Saxons in the sixth century. They wore it as a badge on that occasion by an order of St. David.

But in warmer countries onions and garlic are much more important, where they have flavoured almost every dish since the days of Nestor's banquet to Machaon in Asia, and of the Emperor Nero in Italy, until our own days.

But the subject is so inexhaustible, depending as it does upon man's powers of invention and his tendency to weird superstitions, that we must close this chapter and also the book.

And we will end by asking the reader to think sometimes of all these many and various ways in which plants help and interest man.

It is not merely because our life depends upon them. Everything that we eat has been produced by plant life and plant work.

Tea, coffee, cocoa, and wine are pleasant because plants have produced some essence which is found useful and agreeable by mankind. Even water would be tasteless and unwholesome were it not for the minute diatoms and other microscopic vegetables in it.

But those who take an interest in flowers and leaves *for themselves*, find that they need never spend a dull hour in the country. There is so much to see and to find out, even in the commonest weed or the tiniest floweret.

But it is necessary to sympathize with them, to try to look at things from their point of view, and not merely from an artistic or collector's standpoint.

The romance of plant life then becomes a fascinating and engrossing pursuit. But however long one studies it, the knowledge that the wisest naturalist can ever attain to must remain a negligible quantity compared with what he does not know.

Suppose a mouse happened to stray into the office of the editor of the *Times,* he might boast to his fellow-mice of his knowledge of the "higher journalism," but his opinions would not really be of very great value on the subject.

However hard we study, and however much we observe and reflect upon the working of this great world of Nature, we really cannot expect to know more relatively than that little mouse.

In fact, the more we think, the more humble men of heart we become, and the greater also should be our reverence for the Creator of this wonderful universe.